MATERIALS AND PROCESSES
FOR SUBMICRON TECHNOLOGIES

EUROPEAN MATERIALS RESEARCH SOCIETY SYMPOSIA PROCEEDINGS

MATERIALS AND PROCESSES FOR SUBMICRON TECHNOLOGIES

PROCEEDINGS OF SYMPOSIUM N ON
MATERIALS AND PROCESSES FOR SUBMICRON TECHNOLOGIES
OF THE E-MRS 1998 SPRING CONFERENCE

STRASBOURG, FRANCE, 16-19 JUNE 1998

Edited by

J.M. MARTINEZ-DUART
Depto. Fisica Aplicada, Universidad Cantoblanco, Madrid, Spain

R. MADAR
LMPG-ENSPG, Saint Martin d'Hères, France

R.A. LEVY
New Jersey Institute of Technology, Newark, NJ, USA

1999

ELSEVIER
AMSTERDAM - LAUSANNE - NEW YORK - OXFORD - SHANNON - SINGAPORE - TOKYO

ELSEVIER SCIENCE Ltd
The Boulevard, Langford Lane
Kidlington, Oxford OX5 1GB, UK

ISBN 0-08-043617-X
Reprinted from: Solid State Electronics, vol. 43/6

Library of Congress Cataloging in Publication Data
A catalog record from the Library of Congress has been applied for.

British Library Cataloguing in Publication Data
A catalogue record from the British Library has been applied for.

⊗ The paper used in this publication meets the requirements of ANSI/NISO Z39.48-1992 (Permanence of Paper).
Transferred to digital print on demand, 2006
Printed and bound by CPI Antony Rowe, Eastbourne

SOLID-STATE ELECTRONICS

VOLUME 43 NUMBER 6

JUNE 1998

CONTENTS

**Papers presented at the European Materials Research Society 1998 Meeting.
Symposium N: Materials and Processes for Submicron Technologies**

[continued overleaf

SOLID-STATE ELECTRONICS

VOLUME 43 NUMBER 6   JUNE 1998

CONTENTS—continued]

PERGAMON

Solid-State Electronics 43 (1999) 1001–1002

SOLID-STATE ELECTRONICS

Foreword

It is well known that advances in microelectronic devices are directly related to the reduction in size of the features of the devices down to the submicron size (close to 0,1 μm). When the electronic materials are patterned to these small sizes, their physico-chemical properties show many new aspects (interdiffusion, electromigration, etc.), many of them not well know yet. For this reason, the European Materials Society decided to hold a Symposium entitled *Materials and Processes for Submicron Technologies* in June 16–19, 98, within the yearly E-MRS Spring Meeting in Strasbourg, France. The articles presented in this volume, 28 in number, are representative of the 40 papers accepted for the last E-MRS Meeting (Symposium N). The accepted articles were classified in several sessions: metallizations, lithography, etc.

Probably the main problem encountered when reducing the devices to the submicron size is related to metallic contacts and interconnects, and this fact was reflected by the large number of papers, about 40%, presented dealing with these topics. As emphasized by R. Liu of Bell Laboratories, the cost of interconnect is estimated at about 50% of the total cost making a chip. In addition the complexity of interconnect increases with the number of devices and the cost of making a connection is much higher off the chip than on the chip; consequently cost competitiveness derives system functions from the PC board to the silicon chip. Many of the papers on contacts and interconnects dealt with the last breakthroughs, appearing in the 1990's, such as reactively sputtered TiN widely used as a barrier/glue layer, use of CVD Tungsten plugs in contact and vias, substitution of Al by Cu which is 40% more conductive, and the introduction of global planarization by chemical mechanical polishing.

The need for new dielectric materials and processes is two fold, depending on whether the dielectric is used as the gate in MOS devices, or as interlevel insulator for multiple metallization levels. In the first case, a situation has been reached in the technologies approaching 0,1 μm, in which the thickness of the silicon oxide gate insulator is so small that produces current leakage and dielectric breakdown. Therefore, there is a need for high K materials (the paper presented on Ta_2O_5 and TiO_2 is along this line) to maintain enough capacitive coupling between the gate electrode and the inversion channel in the semiconductor. On the contrary, the problem with the interlevel dielectrics needs of materials with K as low as possible, since at a feature size of 0,1 μm, the RC interconnect response time is too high, close to 10 ps. In this sense, interesting research on nanoporous silica, fluorinated dielectrics, low K polymers, etc. was discussed in the Meeting. It was also pointed out that compatibility with low resistance Cu metallization schemes could be an extra advantage.

As it is known, further down scaling of device dimensions to 0,1 μm features depends on advances in new lithography techniques, such x-rays or e-beam, since it is recognized that optical lithography, even with the use of phase contrast masks and deep UV sources, cannot be further extended. In this respect there is an emerging area of research in new masking and resist materials and technologies. But perhaps the main emphasis in the discussion was given to the new nanometer scale lithography based on Scanning Tunneling Microscopy (STM) and Atomic Force Microscopy (AFM), which have demonstrated good capabilities for atomic-level manipulation and local modification of surfaces, using different approaches such as selective oxidation.

Since the minimum feature size of the novel devices is approaching the wavelength of electron and holes, quantum effects are becoming more and more important, and the study of two-dimensional quantum systems (2D) becomes necessary. In the past, quantum effects had already been observed in MOSFET'S; in fact, the integer quantum hall effect was discovered in this system. At present, the combination of semiconductor heterostructures and lithograpgy is producing new devices such as quantum well lasers, high electron mobility transistors (HEMT), etc. based on low dimensional semiconductors. As a result, for terahertz electronics to become feasible, it is necessary that the active length of the devices approaches 0,1 μm. The real challenge, at present, is the possibility of tailoring a material in 3D on the scale around 100 nm. These so-called quantum dots are expected to improve the behaviour of lasers in the future.

Most of the papers presented on nanomaterials referred to porous silicon and to semiconductor quantum dots. These are usually made from inorganic semiconductor materials such as Si, CdSe, InP, etc. Quantum dots can be prepared by colloidal chemistry techniques and their size is between 10 and 100 nm, i.e. smaller than the structures manufactured by IC lithographic techniques. In order to explain their properties, it is necessary to apply quantum mechanics. For this reason, this field is very exciting to both theoretical physicists and experimental spectroscopists. One of the most interesting nanomaterials is porous silicon, a material to which several papers were dedicated. Since the discovery by Canham in 1990 of a very strong visible luminescence in porous silicon, which was attributed to quantum size effects, a lot of research has been conducted on this material. Electroluminescent porous

Si devices have been constructed in the last years, and an electroluminiscence efficiency greater than 10^{-3} has been reported for porous silicon LED'S.

As Editor of the papers presented at this Symposium, I would like to thank Drs. R. Levy (New Jersey Institute of Technology) and R. Madar (I.N. Polytechnic, Grenoble) who acted as Symposium organizers, for their helpful suggestions and support. I am also indebted to Drs. Liu (Bell Labs), Torres (LETI, Grenoble) and V. Parkhutik, (UPV, Spain) for their very interesting invited conferenees, where the future trends of nanoelectronics where emphasized, and for their contribution to make the Symposium very profitable and friendly. I would like also to recognize the help received from the very efficient Elsevier staff during the days of the Symposium. Finally, I would like to thank P. Cortés for her help in word processing tasks and elaboration of the program.

J.M. Martinez-Duart
Invited Editor,
E-MRS Vicepresident,
Spain

PERGAMON

Solid-State Electronics 43 (1999) 1003–1009

SOLID-STATE ELECTRONICS

Interconnect technology trend for microelectronics

Ruichen Liu*, Chien-Shing Pai, Emilio Martinez

Bell Laboratories, Lucent Technologies, 700 Mountain Avenue, Murray Hill, NJ 07974, USA

Received 8 July 1998; accepted 18 October 1998

Abstract

In the past 30 years interconnect evolved from a single layer of Al deposited by evaporation to multiple levels of sandwiched Ti/Al–Cu/TiN metal layers deposited by sophisticated magnetron sputtering and connected by Al or W plugs through via holes. Dielectric insulator evolved from low-pressure chemical vapor deposited (LPCVD) SiO_2 film to high density plasma enhanced (HDP CVD) low temperature film for multilevel structures. The development of chemical mechanical polishing (CMP) allowed the building of seemingly unlimited number of levels of interconnects. Yet, with such astonishing advancement the progress of interconnect technology seems to fall short of the expectations from high performance circuits and at each technology node interconnect delays become a higher percentage of the total circuit delay. New materials such as Cu and low-K dielectric insulators are being developed to relieve the performance bottleneck and new structures such as dual-damascene are being investigated to simplify interconnect processes. These measures, assuming their success, will slow the cost escalation but will not provide long term solutions, since materials limits for lower resistivity and dielectric constant will be reached and further reduction will not be feasible. Is a revolution for interconnect technology imminently pending and what will it be? © 1999 Published by Elsevier Science Ltd. All rights reserved.

1. Introduction

In the past 30 years interconnect, which is the passive wiring connecting active devices, evolved from a single layer of Al deposited by evaporation to multiple levels of sandwiched Ti/Al–Cu/TiN metal layers deposited by sophisticated magnetron sputtering or CVD process [1–3]. Dielectric insulator evolved from low-pressure chemical vapor deposited (LPCVD) SiO_2 film to high density plasma enhanced (HDP CVD) low temperature film for multilevel structures [4,5]. The development of chemical mechanical polishing (CMP) allowed the building of seemingly unlimited number of levels of interconnects. Yet, with such astonishing

advances the progress of interconnect technology still falls short of the demand from high performance circuits and at each technology node interconnect delay becomes a higher percentage of the total circuit delay. On the other hand, the cost of metal and insulator deposition tools have escalated from US$50,000 to more than US$2,000,000 a piece and at the 180 nm node the cost of interconnect is estimated at greater than 50% of the total cost of making a chip. Is a revolution for interconnect technology imminently pending and what will it be?

2. The driving force for interconnect: complexity and cost

Modern integrated circuits consist of tens to hundreds of millions of devices to achieve complex functions. Interconnects are the wires that provide the

----

* Corresponding author. Tel.: +1-908-582-4414; fax: +1-908-582-2793.

E-mail address: rcl@lucent.com (R. Liu)

0038-1101/99/$ - see front matter © 1999 Published by Elsevier Science Ltd. All rights reserved.
PII: S0038-1101(99)00015-5

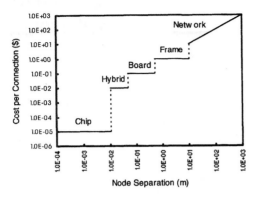

Fig. 1. The cost of making a connection (after Ref. [6]).

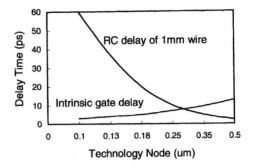

Fig. 3. Intrinsic gate delay ($C_g V/I$) and interconnect RC delay at minimum design rules of each node.

power, the ground, inputs and outputs and timing signals to these devices. The complexity of interconnect thus increases with the number of devices. For random logic circuits, the number of interconnect segments increases linearly with the number of devices [6] and can be expressed approximately as:

$$N_{interconnect} \sim k \times N_{gate} \{ N_{fan\text{-}out} / (1 + N_{fan\text{-}out}) \}$$

where $N_{fan\text{-}out}$ is the average number of fan outs of a gate and N_{gate} is the number of (logic) gates and k is an empirical constant. In general, the number of interconnect segments in an integrated circuit is approximately 10–100 times the number of devices.

However, as shown in Fig. 1, the cost of making a connection is much higher off the chip than on the chip [7] and consequently cost competitiveness drives system functions from the PC board to the silicon chip. This increase in complexity is achieved by integrating more devices and functions which results in both larger chip sizes and more complex interconnect structures. Defect density, on the other hand, controls

the yield (and thus the cost) of the silicon chip. As shown in Fig. 2, the higher degree of on-chip integration is only accomplished through a continuous effort of defect reduction. At any given time the achievable defect level is dictated by the processing equipment available at the time; thus, beyond certain chip size and complexity the cost due to yield loss escalates.

When the complexity of the chip and thus the complexity of the interconnect increases, the area used for interconnect increases rapidly. The result is an inefficient use of silicon area which leads to larger chip size and higher cost. Multiple levels of wiring stacked on one another, connected through via holes similar to that in PC boards, are developed to achieve more efficient use of the silicon area. In an optimally designed circuit, densely packed devices and several densely packed wiring layers coexist. The saving in more efficiently using the silicon area is partially offset by the cost of adding more layers of wiring. However, as shown in Table 1, the benefit of using multilevel wiring greatly offset the cost of additional processing, as far as good yield can be achieved. When systems become more complex and more system functions are transferred to the chip, the complexity of integrated circuits on silicon continues to increase and more level of wiring is needed to keep up with efficient use of silicon area. Therefore, even if processing and design technologies can keep up with the increase in complexity, the cost of interconnect continues to increase. (However, the cost of the system continues to decrease.)

Increased chip size inevitably translates into longer global interconnect wires. As shown in Table 2, RC delay can be scaled only if the wire length is scaled with design rules. When the chip size increases the RC delay on the global interconnect increases. As shown in Fig. 3, the interconnect delay of a 1 mm long wire at minimum design rules is only a fraction of the gate delay at 350 nm node but becomes comparable to the gate delay at 250 nm gate length. At 180 nm gate

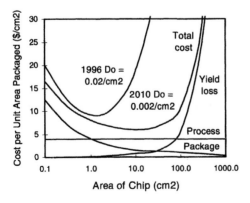

Fig. 2. Cost of fabrication as a function of chip area.

Table 1
Cost and yield analysis of adding more levels of interconnect. A Poisson distribution of defect is used for yield estimates

Number of metal levels	Relative cost	Die area shrinkage (%)	Relative yield (%)	Relative cost per good die
1	1.00	0	90	1.00
2	1.15	10	82	1.15
		30	86	0.87
		50	90	0.58
		70	94	0.33
3	1.30	20	78	1.22
		30	80	1.05
		40	83	0.86
		50	85	0.69
		70	91	0.39
4	1.45	40	78	1.02
		50	81	0.82
		70	88	0.45

length, the interconnect delay of a 1 mm wire is estimated to be about four times that of the gate delay.

Comparing the RC delay of a (semi-) global interconnect line with device gate delay, however, is unfair since large capacitive load is usually driven by wide transistors. An alternative measure is to use the clock frequency. Table 3 shows the clock period and the RC delay of 1 and 2 mm semi-global lines and a 10 mm global line at several technology nodes. The clock frequencies for 250 through 70 nm nodes are based on the 1997 US National Technology Roadmap [8], and the clock frequency for the 350 nm node is based on the 1994 roadmap [9]. The RC delay of a 1 mm semi-global wire will exceed 10% of the clock period at 100 nm node and that for a 2 mm wire will happen earlier at 180 nm node. Consequently, semi-global wires longer than 2 mm cannot be used at minimum design rules at 180 nm node and extra area must be sacrificed to allow the use of wider wires. It is also easy to see that the long RC delay prohibits the use of minimum design rules for global wires.

3. Interconnect trend: innovation in the past three decades

Table 4 summarizes the evolution of interconnect. At the very center of this tremendous progress is the introduction of several new materials and processes. Without these innovations modern planar multilevel interconnect would be impossible.

In the 1970's polysilicon technology and a single layer of Al wiring were almost universally used for all VLSI circuits. Interestingly, because of the maturity of the polysilicon technology [10], multilayer polysilicon was used to facilitate the cross over of wires. The high resistivity of polysilicon was tolerated because the speed of the system was much lower then and the signal delay on wide gate was not limiting the performance.

The 1980's marked the advance of 2–3 levels of metal and of the use of silicides. The use of silicides over polysilicon [11] allowed the circuit designers to use long polysilicon (polycide) lines up to a hundred microns. This efficient use of polysilicon allowed much

Table 2
Scaling of global interconnect ($S \sim 1.4$ per generation). Realistic scaling gives RC between S^2 and S^4

	Chip size scales with design rules	Chip size stays unchanged	Chip size scales inversely with design rules
R (global)	S	S^2	S^3
C (global)	S^{-1}	1	S
RC	1	S^2	S^4

Table 3
Clock frequency for high performance ASIC based on 1997 and 1994 SIA roadmap [8,9], compared to the RC delay of 1, 2 and 10 mm wires at minimum design rules

Node (nm)	350	250	180	130	100	70
Clock frequency (MHz)	300	750	1200	1600	2000	2500
Power supply (V)	3.3	2.5	1.8	1.5	1.2	0.9
Clock period (ns)	3.3	1.3	0.83	0.62	0.50	0.40
1 mm wire RC (Al/SiO_2) (ns)	0.005	0.01	0.02	0.04	0.08	0.16
2 mm wire RC (Al/SiO_2) (ns)	0.02	0.04	0.08	0.16	0.32	0.64
10 mm wire RC (Al/SiO_2) (ns)	0.5	1.0	2.0	4.0	8.0	16

design flexibility in the local level and effectively delayed the need for multilevel metals. In fact, up to about 4 Mb many DRAM products used only a single metal layer with several layers of polysilicon. In random logic and in microprocessor, however, the complexity of the circuit demanded more complex interconnect and multilevel metallization was introduced into these products.

Two advances in processing technology allowed the development of multilevel metallization, plasma enhanced CVD deposition of dielectric material between metal layers (ILD) and planarization of the ILD. Plasma enhanced CVD process allowed the deposition of ILD at low temperature compatible with Al wiring. Planarization processes produced locally smooth ILD surfaces for a base to deposition the next layer metal without breakage. However, these local planarization processes did not remove the total topography built up on a chip. After the first one or two levels of metal, the total topography on a chip became larger than the allowed depth of focus (DoF) for optical lithography. Until a global planarization process was invented the number of metal levels usable was limited to below three.

The late 1980's and the early 1990's were marked with three breakthrough processes, reactively sputtered TiN, CVD W plug and chemical mechanical polishing

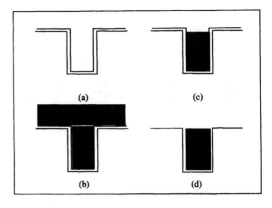

Fig. 4. CVD W plug process: (a) window with TiN liner, (b) CVD W deposition and (c) RSE etch to remove W from TiN surface, leaving a W plug. (d) An alternative method to form the W is to remove both W and TiN by CMP.

(CMP). Reactively sputtered TiN allowed the use of TiN cladding on Al wiring and thus greatly increased the electromigration and stress migration resistance of Al wiring. In addition, TiN was widely used as a barrier/glue layer for CVD W. The CVD W plug process is shown in Fig. 4. Since CVD W has high resistivity and rough morphology, the W film on the surface is removed, leaving only a W plug in the contact window or vias. After the W plug formation, AlCu and TiN are deposited over the W plug as the new metal layer. The use of a plug in contacts and vias allows the stacking of vias which further reduces the area used for interconnect.

Chemical mechanical polishing was first introduced for polishing plate glass to remove distortions in 1920's [12]. The process has been adopted for polishing silicon wafers after the wafers were sawed from ingots. The application to ILD planarization was first introduced at IBM [13]. Unlike other planarization processes, CMP provides a global planarization because

Table 4
Interconnect technology milestones

	Design rules (µm)	MPU speed (MHz)	Interconnect technology	Enabling process/technology
1970's	10–2	< 10	polysilicon, 1 metal level	polysilicon
Early 1980's	2–1.3	20	polycide, 2 metal levels	polycide PECVD–SiO_2
Late 1980's	0.8–0.6	80	salicide, 3 metal levels	self-aligned silicide, reactive TiN, CVD W plug, planarization
Early 1990's	0.5–0.35	200	salicide, TiN local interconnect, 4–5 metal levels	CMP, HDP ILD
Late 1990's	0.25–0.18	400	salicide, W local interconnect, 5–6 metal levels	low-K ILD, Cu metal, damascene

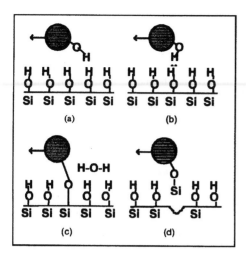

Fig. 5. Mechanism of CMP: (a) oxides form hydroxyls in aqueous solution, (b) forming of hydrogen bond between the slurry particle and the surface, (c) a Si–O bond is formed by releasing a H_2O and (d) the Si–Si bond is broken when the slurry particle moves away.

the polishing is referenced to a large, flat polishing pad and platen. Fig. 5 shows the basic mechanism for materials removal. As can be seen, in molecular level, the polishing is accomplished by chemical reactions between the surface and the polishing slurry. The mechanical structure of the polishing pad, which controls the local pressure, determines the resulting planarity.

The global nature of CMP provides two significant advantages: (1) a relieve on optical lithography DoF and (2) a stop of topography and defect propagation. The latter is especially important, because they allow the building of unlimited levels of interconnect with cost as essentially the only criterion. As shown in Fig.

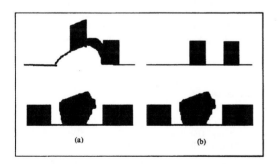

Fig. 6. CMP prevents the propagation of defects: (a) nonfatal defects in lower level causes shorts in upper level metal, (b) CMP stops the propagation.

Table 5
Interconnect levels predicted by 1997 SIA National Technology Roadmap. Note that the use of low-*K* dielectric ILD and Cu interconnect is already factored into this table

Year	1997	1999	2003	2006	2009	2012
Tech node (nm)	250	180	130	100	70	50
Metal level	6	6–7	7	7–8	8–9	9

6, a non-fatal defect in a lower level of interconnect can be exaggerated in subsequent processes to cause a fatal short in the level above. CMP stops the propagation of such defects, as shown in Fig. 6(b), by arresting the growth of non-fatal defects. When defects from the lower level are allowed to propagate to upper levels the yield of upper level interconnects worsens progressively. Since the total yield is the product of the yield of individual levels a defect in lower level executes a high yield loss when propagated.

The advances in high density plasma (HDP) enhanced CVD dielectric process, the use of CVD W and Al plugs in contacts and vias and the introduction of global planarization by CMP have allowed the use of multilevel metallization to fabricate highly integrated circuits today. At the very center of these advances are innovations in new materials and processes. A huge amount of research and development work has been done to make these enabling technologies practical, as witnessed by the large number of publications in literature (for a collection of references, see Ref. [14]).

4. Interconnect trend: new innovations in the near future

High density, planar, multilevel interconnect allows the building of sophisticated circuits on silicon and to capitalize on the reduction in cost more functions are moved from board to chip. Cost, however, is not the only driving force. The demand for more sophisticated software and even faster data transmission for multimedia and network applications dictates higher circuit speed, as shown in Table 3. To achieve high speed and to fully utilize silicon area more levels of interconnect is needed at each new technology node. Table 5 shows the projected number of interconnect levels for each technology node based on the SIA 1997 national technology roadmap (NTR) [8]. At 180 nm design rule microprocessors are expected to have at least six levels of interconnect. The cost of interconnects becomes more than 50% of the total wafer cost. Therefore, although the development of multilevel wiring allows the building of complex circuits, the capital investment

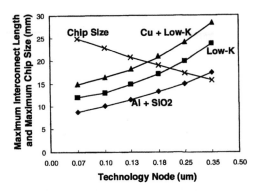

Fig. 7. Maximum interconnect length at each technology node. The maximum length is calculated using clock frequencies given by 1997 SIA NTR shown in Table 3. $K = 2.5$ for the low-K dielectric (from Ref. [15]).

required for new fabrication lines as well as the cost of interconnect continue to escalate.

An alternative to adding more levels of interconnect is to use lower resistance wires and/or lower dielectric constant ILD. Fig. 7 shows the maximum wiring length allowed at each technology node [15]. To replace Al with Cu which has about a 40% higher conductivity and/or to replace SiO_2 with a low-K dielectric ILD (K about 2.5) can significantly increase the maximum wiring length. Consequently, the use of Cu wiring and low-K ILD can improve the performance of a circuit without adding more levels of wiring.

Cu can be deposited by a CVD process [16], by electroplating or electroless plating [17] or by PVD processes [18]. Cu, however, imposes four major obstacles: (1) Cu can easily diffuse through SiO_2 and into Si causing device leakage and gate oxide yield and reliability problems, (2) Cu does not reduce SiO_2 and thus does not adhere well, (3) Cu does not have a non-porous oxide such as Al_2O_3 and thus is vulnerable to corrosion and (4) Cu has no inorganic volatile compound and thus is difficult to etch by conventional RIE processes. At this time, barrier and glue materials are still being developed to contain Cu diffusion and to improve adhesion. Damascene process, for which trenches are etched into the ILD and then filled with metal, is adopted instead of Cu RIE. Multilevel Cu damascene processes have been recently demonstrated by both electroplated Cu [19,20] and by CVD Cu [21,22]. There seems to be no fundamental barrier for a mature Cu technology for 0.18 µm circuits, however, the development of manufacturing-worthy Cu processing equipment remains a major task.

Organic low-K ILD materials generally suffer from low mechanical strength and stability, low thermal stability and poor thermal conductivity. Several spin-on and vapor deposited low-K polymers have been developed [23–25]. Successful integration of organic low-K materials with Al metallization has so far been only limited, since both Al and W plug processes require temperatures above 400°C. Reliability issues of low-K ILD and the impact of Cu wiring and low-K ILD still need to be addressed. However, for high performance circuits, the use of Cu and low-K ILD will allow a temporary pause of escalating interconnect cost.

5. Beyond new materials

Beyond Cu the improvement in electric conductivity for any new metal is insignificant (except for high T_c superconductors which have poor current density thresholds). Similarly, the dielectric constant of low-K ILD is unlikely to go below 1.5. Thus the potential benefit of new materials is very limited once Cu and low-K ILD with $K \sim 2$ are used.

As can be seen in Fig. 7, at 180 nm node the maximum wiring length using Al and SiO_2 is already substantially smaller than the maximum chip size. Cu and low-K ILD can help stretching the crossover point by about two generations. After that either the clock frequency will start to saturate, or the chip size will stop increasing with the consequence of saturating the cost per function. Both will lead to deviation from the Moore's law and the consequent stagnation in future market expansion.

New structures, such as coaxial interconnect and optical interconnect, may find applications in specialized products. General solutions, however, are still lacking. It seems that finally evolutionary improvement has reached the end of road.

However, there are no revolutionary solutions in sight either!

The power of innovation, however, should never be underestimated. The forces of the US$135 billion dollar semiconductor industry and the nearly one trillion dollar computer, communication and electronics industry will most likely continue to create new solutions. In the end, perhaps, the speed of light is the only limitation.

References

[1] Glang R. Vacuum evaporation. In: Maissel LI, Glang R, editors. Handbook of thin film technology. New York: McGraw-Hill, 1970. p. 1–300.

[2] Rossnagal SM. Ionized magnetron sputtering (I-PVD) for lining and filling trenches and vias at room temperature. In: Proceedings 1995 International VLSI Interconnection Conference, 1995. p. 576.

[3] Kang SB, Chae YS, Yoon MY, Leem HS, Park CS, Lee SI, Lee MY. Low temperature processing of conformal TiN by ACVD (advanced chemical vapor deposition) for multilevel metallization in high density ULSI Devices. In: Proceedings 1998 International Interconnect Technology Conference, 1998. p. 102.

[4] Adams AC, Capio CD. The deposition of silicon dioxide films at low pressure. J Electrochem Soc 1979;126:1042.

[5] Chebi R, Mittal S. A manufacturable ILD gap fill process with biased ECR CVD. In: Proceedings of the 8th International IEEE VLSI Multilevel Interconnection Conference, 1991. p. 61.

[6] Donath WE. Placement and average interconnections lengths of computer logic. IEEE Trans Circ Syst 1979;26(4):272.

[7] Mayo JS. Materials for information and communication. In: Scientific American, 1986. p. 59.

[8] The national technology roadmap for semiconductors, Semiconductor Industry Association, 4300 Stevens Creek Boulevard, Suite 271, San Jose, CA 95129, 1997. p. 16.

[9] The national technology roadmap for semiconductors, Semiconductor Industry Association, 4300 Stevens Creek Boulevard, Suite 271, San Jose, CA 95129, 1994. p. 41.

[10] Vadasz LL, Grove AS, Rowe TA, Moore GE. Silicon gate technology. IEEE Spectrum 1969;6:28.

[11] Murarka SP, Brown DB, Sinha AK, Levinstein HJ. Refractory silicides of Ti and Ta for low-resistivity gates and interconnection. IEEE Trans Electron Devices 1980;27:1425.

[12] Preston FW. The theory and design of plate glass polishing machines. J Soc Glass Technol 1927;11:27.

[13] Davari B, Koburger CW, Schulz R, Warnock JD, Furukawa T, Jost M, Taur Y, Schwittek WG, DeBrosse JK, Kerbaugh ML, Mauer JL. In: A new planarization technique using a combination of RIE and chemical mechanical polish (CMP). 1989 IEDM Technical Digest, 1989. p. 61.

[14] Liu R. Metallization. In: Chang CY, Sze SM, editors. ULSI Technology. New York: McGraw-Hill, 1996. p. 371–471.

[15] Rahmat K, Nakagawa OS, Oh S-Y, Moll J, Lynch WT. In: A scaling scheme for interconnect in deep-submicron processes. 1995 IEDM Technical Digest, 1995. p. 245.

[16] Cho JSH, Kang HK, Asano I, Wang SS. In: CVD Cu interconnection for ULSI. 1992 IEDM Technical Digest, 1992. p. 297.

[17] Pai PL, Ting CH. Copper as the future interconnection materials. In: Proceedings of the 6th International IEEE VLSI Multilevel Interconnection Conference, 1989. p. 258.

[18] Abe K, Harada Y, Onoda H. Sub-half micron copper interconnects using reflow of sputtered copper films. In: Proceedings of 13th International VLSI Multilevel Interconnection Conference, 1995. p. 308.

[19] Edelstein D, Heindenrich J, Goldblatt R, Cote W, Uzoh C, Lustig N, Roper P, McDevitt T, Motsiff W, Simon A, Dukovic J, Wachnik R, Rathore H, Schutz R, Su L, Luce S, Slattery J. In: Full copper wiring in a sub-0.25 µm CMOS ULSI technology. 1997 IEDM Technical Digest, 1997. p. 773.

[20] Heidenreich J, Edelstein D, Goldblatt R, Cote W, Uzoh C, Lustig N, McDevitt T, Stamper A, Simon A, Dukovic J, Andriacacos P, Washnik R, Rathore H, Kasetos T, McLaughlin P, Luce S, Slattery J. Copper dual damascene for sub-0.25 µm CMOS. In: Proceedings 1998 Interconnect Interconnect Technology Conference, 1998. p. 151.

[21] Venkatesan S, Gelatos AV, Misra V, Islam R, Smith B, Cope J, Wilson B, Tuttle D, Cardwell R, Yang I, Gilbert PV, Wodruff R, Bajaj R, Das S, Farkas J, Watts D, King C, Crabtree P, Sparks T, Lii T, Simpson C, Jain A, Herrick M, Capasso C, Anderson R, Venkatraman S, Fillipiak S, Fiordalice B, Reid K, Klein J, Weitzman EJ, Kawasaki H, Angyal M, Freeman M, Saaranen T, Tsui P, Bhat N, Hamilton G, Yu Y. In: A high performance 1.8 V, 0.2 µm CMOS technology with copper metallization. 1997 IEDM Technical Digest, 1997. p. 769.

[22] Zhang J, Denning D, Braeckelmann G, Venkatraman R, Fiordalice R, Weitzman E. CVD Cu process integration for sub-0.25 µm technologies. In: Proceedings 1998 Interconnect Technology Conference, 1998. p. 163.

[23] Townsend PH, Martin SJ, Godschlax J, Romer DR, Smith Jr. DW, Castillo D, DeVries R, Buske G, Rondan N, Froelicher S, Marshall J, Shaffer EO, Im J-H. SiLK polymer coating with low-dielectric constant and high thermal stability for ULSI interlayer dielectric. In: Case CB, Kohl P, Kikkawa T, Lee WW, editors. Low-dielectric constant materials, III. Pittsburgh, PA: Materials Research Society, 1997. p. 9.

[24] Rosenmayer T, Hammes J, Bartz J, Spevack P. An oxide cap process for a PTFE-based IC dielectric. In: Proceedings 1998 International Interconnect Technology Conference, 1998. p. 283.

[25] Plano MA, Kumar D, Cleary TJ. The effect of deposition conditions on the properties of vapor-deposited parylene AF-4 films. In: Case CB, Kohl P, Kikkawa T, Lee WW, editors. Low-dielectric constant materials, III. Pittsburgh, PA: Materials Research Society, 1997. p. 213.

PERGAMON

Solid-State Electronics 43 (1999) 1011–1014

**SOLID-STATE
ELECTRONICS**

Factors which determine the orientation of CVD Al films grown on TiN

M. Avinun[a],*, W.D. Kaplan[a], M. Eizenberg[a], T. Guo[b], R. Mosely[b]

[a]*Department of Materials Engineering, Technion-Israel Institute of Technology, 32000 Haifa, Israel*
[b]*Applied Materials, Santa Clara, CA, USA*

Received 17 July 1998; received in revised form 26 October 1998; accepted 17 January 1999

Abstract

CVD Al deposited on a TiN/Ti liner followed by 'warm' PVD Al (Cu) is a promising approach for filling high aspect ratio vias. A highly (111) textured film is important to achieve high electromigration resistivity. The dependence of the Al orientation on the type of the underlying TiN liner and the mechanism responsible for this effect were investigated. HRTEM was used to characterize the structure of the Al grains (after 1 s of deposition), through to a continuous film, as well as the structure of the interface of thick Al films with TiN. The microscopy results helped to explain the X-ray diffraction observations that on PVD TiN the Al film had a stronger (111) texture than on other TiN types. This is attributed to a topotaxial relationship, which forms during the nucleation stage of the CVD Al. © 1999 Elsevier Science Ltd. All rights reserved.

Keywords: CVD; Al; TiN; Texture; Preferred orientation; Topotaxial

1. Introduction

Strict requirements on the technology of interconnections, such as the drive for lower resistivity of the gap filling metal and the use of low-*k* intermetal dielectric materials [1], are imposed due to the reduction in semiconductor-device dimensions (below 0.25 μm).

The low resistivity of Al (compared to W) and the good step coverage, make the CVD Al process a good candidate for this application, when followed by 'warm' PVD Al(Cu) deposition. Furthermore, the low process temperature of CVD Al enables the integration of low-*k*-dielectric materials, i.e. organic materials [2], and this is an important advantage over the Al reflow

process [3]. The introduction of di-methyl aluminum hydride (DMAH) as an organometallic precursor [1,4,5] enabled chemical vapor deposition of high quality Al films. We have found that the properties of the Al layer, as well as the nucleation and growth of the Al film, are strongly influenced by the underlying TiN layer [6]. In the current report we focus on the degree of preferred orientation of the Al film as a function of the TiN type (comparing five TiN processes) and on the mechanism which determines this preferred orientation.

2. Experimental procedures

The nucleation and growth of CVD Al was studied on the stack of Si/SiO$_2$/Ti/TiN. The Ti/TiN liner was deposited in an Endura® cluster tool. Three different PVD processes were used to deposit the Ti film and

---

* Corresponding author. Fax: +972-4-832-1978.
E-mail address: avinun@techunix.technion.ac.il (M. Avinun)

0038-1101/99/$ - see front matter © 1999 Elsevier Science Ltd. All rights reserved.
PII: S0038-1101(99)00016-7

Table 1
Sample description

Symbol	Description
P	LP-Ti/PVD TiN/CVD Al
HP	HP-Ti/HP-PVD TiN/CVD Al
C	LP-Ti/5 nm CVD TiCN/CVD Al
X	LP-Ti/2 × 5 nm CVD TiN-XP/CVD Al
Z	IMP-Ti/2 × 5 nm CVD TiN-XZ/CVD Al

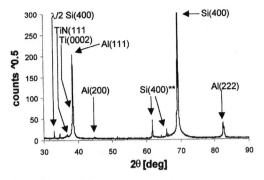

Fig. 1. Typical XRD spectrum of Si/SiO₂/HP-Ti/HP-PVD TiN/300 nm CVD Al. The strong reflection of the Si substrate can be seen for some wavelengths, see Table 2.

five different types of TiN were deposited on the Ti layers (see Table 1).

LP-Ti describes a Ti film, deposited by sputtering at a relatively low power of 8 kW through a collimator with an aspect ratio of 1.5:1. The HP (high power) Ti was deposited in the same chamber, but at a higher power (22 kW) and a lower heater temperature (200°C instead of 250°C). The IMP process is a PVD process utilizing ionized Ti atoms [7].

PVD TiN was sputter deposited using a monolithic Ti target at a power of 6.5 kW and HP-PVD TiN was deposited by deposition at a higher power (20 kW), in the same chamber used for the Ti deposition. The CVD TiCN [8,9] and the CVD TiN-XP (H_2/N_2 plasma treated CVD TiN) [10] were deposited in a lamp heated MOCVD TiN chamber, from the thermal decomposition of TDMAT (tetrakis-dimethyl-amido-titanium) at a processing temperature of 420°C and a pressure of 1.2 Torr. The concept of CVD TiN-XP is maintained for CVD TiN-XZ, but in this case a resistively heated chuck was used instead of lamp heating.

The CVD Al was deposited on top of the Ti/TiN stack. DMAH was used as a precursor with Ar as a carrier gas (400–500 sccm) and the process pressure was 25 Torr. The heater temperature was maintained at 265°C. Under these conditions the deposition rate varied from 4.3 to 6.8 nm/s. In order to study the nucleation and early stages of deposition, we used short deposition times (from 1 s), and in order to study the 'bulk' properties, thick continuous Al films (as thick as 300 nm) were also prepared.

X-ray diffraction (XRD) measurements were carried out (Philips PW-1820) in a Bragg Brentano configuration to determine the preferred orientation of the Al films, using CuK$_\alpha$ radiation ($\lambda = 0.15406$ nm). Cross-sectional transmission electron microscopy (TEM) was carried out to study the microstructure of the Al film and the Al/TiN interface. Both conventional TEM (CTEM, JEOL 2000FX) and high resolution TEM (HRTEM, JEOL 3010 UHR) were used.

3. Results and discussion

XRD was used to evaluate the degree of preferred orientation, using the intensity ratio of the Al (111) and Al (200) peaks [11]. This ratio should be 2.13 for randomly oriented Al (JCPDS 4-0787). A typical XRD spectrum is shown in Fig. 1 and Table 2 represents the corresponding identified peaks. The contribution of the Ti(0002) peak to the Al(111) peak intensity was subtracted after evaluation from the wafer with the lowest

Table 2
Identified XRD peaks

Material and reflection	JCPDS file	d-spacing (nm) (from JCPDS)	d-spacing (nm) (experimental)
TiN (111)	38-1420	0.24492	0.244–0.247[a]
Ti (0002)	5-0682	0.2342	0.23–0.234[a]
Al (111)	4-0787	0.2338	0.2335–0.2339
Al (200)	4-0787	0.2024	0.202
Si (400)	5-0565	0.1357	0.1357[b]
Al (222)	4-0787	0.1169	0.117

[a] Depends on the Ti/TiN type.
[b] This strong reflection was identified in several angles, due to low radiation of CuK$_\beta$, CuK$_\gamma$, W$_{L\alpha1}$, W$_{L\alpha2}$ in addition to the main radiation of CuK$_\alpha$.

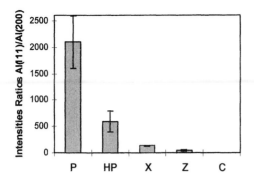

Fig 2. The XRD intensities ratios for Al(111)/Al(200) on different TiN underlayers. The error due to the separation uncertainty of Ti(0002) and Al(111) partial overlap is marked (The intensity ratio for CVD TiCN (C) is 7 ± 2)

amount of Al deposition (1 s of deposition). All the

thick Al films (200–300 nm) were (111) oriented, but the degree of preferred orientation varied for the different TiN underlayers (see Fig. 2).

Fig. 3(a) shows a cross-section TEM micrograph of Al nuclei on PVD TiN after 1 s of deposition. A lattice image of one of the nuclei is given in Fig. 3(b). Even at such an early stage of nucleation/growth, the Al grain has a (111) texture. More than 20 different Al grains were analyzed after 1 s of deposition on various TiN underlayers, and all of the Al grains were oriented with the (111) plane parallel to the interface. The smallest nuclei that were detected were already larger than the underlying TiN grains. The TiN grains are also (111) preferentially oriented and therefore an orientation relationship can exist between at least one TiN grain and the Al nucleus. Thus for some TiN/Al grains a topotaxial [12] relationship exists, including misfit dislocations, as observed in Fig. 3(c). In Fig. 3(c) a misfit dislocation is on the (111) plane and in addition

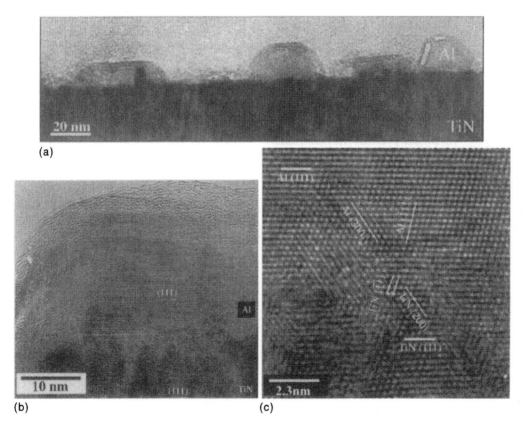

(a)

(b)

(c)

Fig. 3. (a) Bright field cross section of CVD Al nuclei after 1 s of deposition on PVD TiN. (b) Lattice image of a CVD Al nucleus on PVD TiN (zone axis [01$\bar{1}$]). (c) The same micrograph as (b), at a higher magnification, showing a degree of coherency and a misfit dislocation between the Al grain and a TiN grain.

to the general behavior of $(111)_{TiN}\|(111)_{Al}$, the $[200]_{TiN}$ is parallel to $[200]_{Al}$. Therefore, the preferred orientation of the CVD Al film does not evolve during growth due to a fast growth of a specific crystallographic orientation, as may be assumed following the model of Thornton and Van der Drift [13,14], but rather the nucleation itself is textured, due to a topotaxial relationship between the TiN and the Al.

It is well known that TiN deposited by PVD on Ti is (111) textured. In this study the PVD TiN was much thicker film than the HP-PVD TiN film, so it is difficult to say what causes the reduction in the degree of preferred orientation of the TiN film – the change in the process (high power) or the difference in the film thickness. On the other hand, the CVD TiCN film consists of an amorphous matrix, with embedded randomly oriented nanocrystals [8,9]. While the plasma treatment increases the degree of crystallization of the TiN film, the crystals are still randomly oriented and thus the CVD TiN-XP and CVD TiN-XZ have a lower degree of preferred orientation than those deposited by PVD. The XRD spectra and the HRTEM lattice images also indicate that the degree of (111) preferred orientation of the TiN film itself is higher for the PVD and high power PVD TiN samples, than for the CVD TiN with a plasma treatment. Therefore, we conclude that the TiN serves as a template layer for the Al, and determines the orientation of the Al film that grows on it, due to a topotaxial relationship. Moreover, there is no significant influence of the TiN grain size on the degree of preferred orientation of the CVD Al film. High power PVD TiN (HP) had grains with the width of 3 ± 0.2 nm, much smaller grains than those of the PVD TiN (8 ± 1.1 nm) but also smaller grains than 5.8 ± 1.4 nm, which is the CVD TiN-XP (X) width. Yet, the CVD Al which grew on HP was more (111) oriented than on X.

4. Conclusions

An integrated process of CVD Al from thermal decomposition of DMAH on Ti/TiN liners in a cluster tool was investigated. Both TiN and Al grains are (111) textured. The degree of texture of the TiN grains determines the degree of preferred orientation of the CVD Al film, due to a topotaxial relationship which exists between some of the TiN grains and the Al grains at the early stages of deposition. Therefore, the use of a highly (111) textured TiN film improves the degree of (111) preferred orientation of the CVD Al film, which is important for electromigration resistance.

Acknowledgements

MA acknowledges the Israel Ministry of Science for financial support.

References

[1] Dixit GA, Paranjpe A, Hong QZ, Ting LM, Luttmer JD, Havemann RH. IEDM 1995;95:1001.

[2] Zhao B, Wang S-Q, Anderson S, Lam R, Fiebig M, Vasudev PK, Seidel TE. Mater Res Soc Symp Proc 1996;427:415.

[3] Barth HJ. Mater Res Soc Symp Proc 1996;427:253.

[4] Kondoh EK, Ohta T. J Vac Sci Technol A 1995;13:2863.

[5] Matsuhashi H, Gotoh A, Chung JH, Masu K, Tsubouchi K. In: Ellwagner RC, Wang SQ, editors. Advanced metallization and interconnect systems for ULSI applications in 1995. Pittsburgh: MRS, 1995. p. 667.

[6] Avinun M, Barel N, Kaplan WD, Eizenberg M, Naik M, Guo T, Mosely R, Littau K, Zhou S, Chen L. Thin Solid Films 1998;320:67.

[7] Dixit GA, Hsu WY, Konechi AJ, Krishnan S, Luttmer JD, Havemann RH, Forster J, Yao GD, Narasimhan M, Xu Z, Ramaswami S, Chen FS, Nulman J. IEDM Tech Dig 1996;96:357.

[8] Eizenberg M, Littau K, Ghanayem S, Mak A, Maeda Y, Chang M, Sinha AK. Appl Phys Lett 1994;65:2416.

[9] Eizenberg M, Littau K, Ghanayem S, Liao M, Mosely R, Sinha AK. J Vac Sci Technol A 1995;13:590.

[10] Danek M, Liao M, Tseng J, Littau K, Saigal D, Zhang H, Mosely R, Eizenberg M. Appl Phys Lett 1996;68:1015.

[11] Knorr DB, Rodbell KP. J Appl Phys 1996;79:2409.

[12] Sutton AP, Ballufi RW. Interfaces in crystalline materials. Oxford: , 1995.

[13] Thornton JA. Ann Rev Mater Sci 1977;7:239.

[14] Van der Drift. Philips Res Rep 1967;22:267.

PERGAMON

Solid-State Electronics 43 (1999) 1015–1018

SOLID-STATE
ELECTRONICS

Study of Cu contamination during copper integration for subquarter micron technology

P. Motte[a],*, J. Torres[b], J. Palleau[b], F. Tardif[c], H. Bernard[a]

[a]*SGS-Thomson Microelectronics, 850 rue J. Monnet, 38926 Crolles Cedex, France*
[b]*FT/CNET, Chemin du Vieux Chêne, BP98, 38243 Meylan, France*
[c]*CEA/LETI, 17 rue des Martyrs, 38054 Grenoble Cedex 9, France*

Received 9 July 1998; received in revised form 15 November 1998; accepted 16 January 1999

Abstract

Copper contamination in several dielectric deposited on copper-CVD film was investigated. This study aims at integrating copper in a dual damascene structure interconnection for sub-quarter micron technology. A complete contamination profile into the deposited dielectric was available using SIMS, TXRF (total X-ray reflection fluorescence) and LPD–AAS (liquid phase decomposition–atomic absorption spectroscopy) as complementary characterization tools. The Cu contamination profile of the SiN/SiO_2 and SiO_2 structure deposited on copper was given. The PECVD process used for both process cleaning of the chamber and SiOF deposition imply plasma with fluorine and the reactivity of this species with copper was shown to be critical for dielectric contamination and copper contact surface. Eventually, a cleaning solution was investigated to lower the contamination at the surface of the dielectric, the most contaminated part. © 1999 Elsevier Science Ltd. All rights reserved.

1. Introduction

The decrease of ULSI circuit dimensions makes interconnection delay time and current densities limiting parameters for device performances. Due to the low resistivity and good electromigration properties, copper was shown to be a better candidate than aluminum and aluminum-based alloys currently employed. Difficulties in integrating copper at the first metallization level were due to the damages copper could provoke in devices when diffusing into the silicon active zone [1]. Now, several known barrier materials are available to avoid copper diffusion into silicon; problems with deposition, patternability and stability during processing have been extensively studied and first workable solutions are available even at the indus-

trial level [2,3]. However, a lot of work remains to be done to control copper contamination into inter-level and intra-level dielectric materials during the integration process. Effects of copper contamination in dielectric materials are not demonstrated but it is necessary to make it as low as possible to avoid possible leakage currents between metal lines. In this paper a complete study of the contamination problems found during oxide deposition onto copper metallization will be exposed. We will try to highlight the contamination mechanism and evaluate a cleaning solution to lower Cu contamination.

2. Experimental

Copper is integrated in a dual damascene architecture (Fig. 1). The via level dielectric is first deposited, the so called etch stop layer between the via and line

* Corresponding author.

0038-1101/99/$ - see front matter © 1999 Elsevier Science Ltd. All rights reserved.
PII: S0038-1101(99)00017-9

1016   P. Motte et al. / Solid-State Electronics 43 (1999) 1015–1018

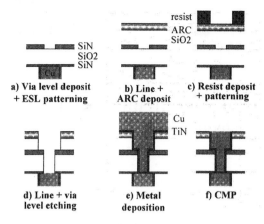

Fig. 1. Dual-damascene structure.

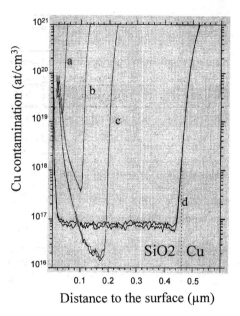

Fig. 2. SIMS profile of Cu into (a) 50 nm, (b) 100 nm, (c) 200 nm and (d) 500 nm thick SiO_2 films, deposited on copper.

level is then deposited and patterned to define a via footprint. After dielectric at the line level, ARC and resist deposition, photolithography of the line footprint in the resist and line and via etching, the metallization is deposited and the Cu metal excess removed by CMP. The dielectric is deposited by plasma assisted chemical vapor deposition (PECVD), at 400°C, in a cluster tool of Applied Materials. The gas deposition processes are SiH_4, N_2 and NH_3 for SiN deposition, SiH_4 and N_2O for SiO_2 deposition and with the addition of CF_4 for fluorine doped oxide (SiOF). The SiOF process was optimized to obtain a 3 at% F containing layer, stable in wet atmosphere (85% TH, 80°C, 24 h aging). For dielectric layers thicker than 5000 Å, a chamber cleaning based on CF_4 and N_2O plasma was necessary between each deposition process. To obtain a complete contamination profile of the dielectric, three complementary characterization tools were used. SIMS (secondary ion mass spectrometry) gives information about the Cu contamination in volume. Surface contamination was studied by TXRF (total X-ray reflection fluorescence) and LPD–AAS (liquid phase decomposition–atomic absorption spectroscopy). The last method consists of etching the surface dielectric in a solution that will then be analyzed by AAS.

3. Results and discussion

We will present first the contamination profile observed in the SiO_2/SiN and SiO_2 layers. We will highlight the role of F contamination by focusing on SiOF deposition on copper. A surface cleaning solution will then be investigated.

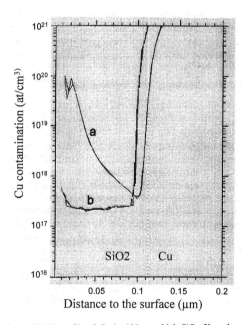

Fig. 3. SIMS profile of Cu in 100 nm thick SiO_2 films deposited on copper, (a) as deposited and (b) after annealing.

Cu contamination results in at/cm3		Dielectric thickness			
		50nm	100nm	200nm	500nm
SiO2/Cu, [Cu]$_{SiO2}$ as deposited	interface	5.10^{19}	4.10^{17}	2.10^{16}	8.10^{16}
	surface	$\approx 10^{20}$	$\approx 10^{20}$	$\approx 10^{20}$	$\approx 10^{20}$
SiO2/Cu annealed	bulk	1.10^{18}	3.10^{17}	2.10^{17}	$1.5.10^{17}$
SiO2/SiN/Cu, [Cu]$_{SiO2}$ [Cu]$_{SiN} \approx 10^{20}$	SiO2/SiN interface	2.10^{17}	8.10^{16}	3.10^{16}	
	surface	$\approx 10^{20}$	$\approx 10^{20}$	$\approx 10^{20}$	
SiOF/SiO2/Cu, annealed	[Cu]$_{SiOF}$ bulk				2.10^{16}

Fig. 4. Summary of SIMS Cu contamination results, for several dielectric structure and thickness.

3.1. Contamination profile in SiN and SiO₂

Barrier films of either conductive or dielectric materials are generally introduced between the Cu metallisation and SiO$_2$. A thin SiN film previous to SiO$_2$ deposition is currently used as a diffusion barrier layer [4] and acts also as an etch stop layer (Fig. 1a). In fact, an unexpected high level of contamination was observed by SIMS into SiN (Fig. 2). A very sharp interface is noted between SiO$_2$ and SiN (10^{20} Cu at/cm^3 into SiN), which shows the difference between these materials for the diffusion of the contaminating Cu species.

The dielectric contamination into SiO$_2$ was then studied as a function of the layer thickness deposited on Cu film (SIMS results shown in Figs. 2 and 3). The general aspect of the Cu profile contamination found in SiO$_2$ was a very high contamination at the free surface of the SiO$_2$ (10_{20} at/cm^3) compared to the bulk, for which the contamination shows a minimum at the SiO$_2$/CU interface.

Moreover, the bulk contamination is shown to dramatically decrease with increasing oxide thickness. On the other hand, after annealing for 1 h at 450°C under vacuum (Fig. 4), the contamination over the volume observed for as deposited samples decreases down to a constant value of about 2×10^{17} at/cm^3, as shown in Fig. 2. This value nearly corresponds to the same final contamination value for any SiO$_2$ thickness after annealing. Finally, in order to obtain more precise information about the surface contamination, LPD–AAS analysis was carried out which gives a result for 500 nm thick SiO$_2$, a surface peak of 10^{18} Cu at/cm^3 and a contamination limited to the first 50 nm.

An interpretation of all these observations can be proposed, i.e. the formation of a contaminant species at the interface with copper which can diffuse out rapidly through the dielectric layer toward the surface all along the deposition process carried out at 400°C, as well as during annealing. We have no information

on the impact of such a contamination on dielectric properties.

As [F] contamination was observed in all the SIMS spectra, we believe that CuF$_x$ species could be formed with a potential for such behavior (high vapor pressure at 400°C). To check such an assumption, we performed SiOF deposition on top of Cu metallization. Just after SiOF deposition, the Cu film becomes dark brown colored. During annealing, SiOF and this superficial brown layer peels off and reveals again the bright initial Cu film. The SIMS profile shows a dramatic Cu contamination of about 10^{22} at/cm^3 in the whole SiOF layer.

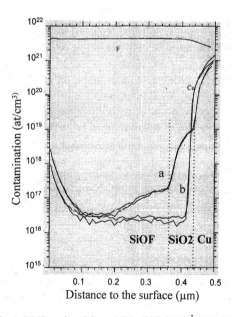

Fig. 5. SIMS profile of Cu and F in SiOF (5000 Å)/SiO$_2$ (500 Å) on Cu, (a) as deposited and (b) after annealing.

Deposition of a 50 nm thin SiO_2 layer before SiOF leads back to a reasonable contamination level (Fig. 5). The SiO_2/SiOF interface is at first well defined and vanishes after annealing; as for SiO_2 deposited on Cu, the contamination in the bulk becomes uniformly low, at 2×10^{16} at/cm³ while surface contamination is two orders of magnitude higher.

SiOF deposition on Cu shows the reactivity of fluorine with copper: only the gas process (CF_4) and plasma power make a difference between the SiO_2 and SiOF process. In other respects, fluorine is continuously present in the deposition chamber due to systematic process cleaning based on F plasma after each dielectric deposition. It was found about a few 10^{19} F at/cm³ in all SiO_2 layers and in SiN. Fluorine may also take part in the contamination and diffusion mechanism of copper observed in both SiO_2 and SiN, in a less dramatic way than in SiOF. Moreover, we found in the SiO_2 layer about 10^{19} Cu at/cm³ in both the SiO_2 (50 nm)/Cu and the $SiOF/SiO_2$ (50 nm)/Cu structures (Fig. 4). For the $SiOF/SiO_2$ (50 nm) structure, the lack of out-diffusion of the Cu species from SiO_2 during the deposition complement could be due to the F gradient imposed during the SiOF process, which should act against diffusion of Cu_xF volatile species toward the surface.

3.2. Dielectric surface cleaning process

Eventually, Cu was shown to be essentially concentrated at the surface of the dielectric, after diffusion over the volume. To diminish the contamination at the dielectric surface, a cleaning solution is then investigated. Its efficiency lies in its ability to remove Cu contaminated oxide and to prevent redeposition of copper by oxidation of the silicon [5]. A study was made on SiO_2 deposited onto Si to evaluate HF 0.1% surface cleaning. This films were contaminated at the surface to a value of about 10^{14} Cu at/cm², by using calibrated contaminating solutions.

TXRF and LPD–AAS were both used as surface characterization tools. The top of the dielectric surface was first etched by some moles of HF and then analyzed by AAS. LPD–AAS is used to measure the copper content in the etched dielectric volume and residual Cu contamination at the surface SiO_2 is controlled by TXRF.

The initial surface contamination is 3.6×10^{13} at/cm³ high (TXRF result), LPD–AAS after 100 nm etching confirmed 3.6×10^{13} at/cm³ and the TXRF analysis gives for the cleaned surface a contamination under 10^{10} at/cm³. By repeating the controlled contamination and etching step on the same dielectric layer, we control the efficiency of Cu removal in the oxide bulk also (TXRF results are 2.10^{10} at/cm³).

Eventually, the HF solution for cleaning was shown to be efficient for both the dielectric surface and bulk that could differ somewhat in composition.

Deposition onto copper is not the only contaminating step. Etching (Fig. 1d) and polishing (Fig. 1f) induce also a physical copper contamination on the surface dielectric through copper resputtering and contact. The cleaning HF solution was here also successfully used for SiO_2 post-CMP cleaning.

4. Conclusion

As a conclusion, fluorine presence in equipment was shown to be responsible for the contamination of dielectric deposited on copper. The contaminating species diffuse easily toward the surface and eventually, while surface contamination is above 10^{18} Cu at/cm³, bulk contamination is found to be about 10^{17} at/cm³ for 5000 Å (and more) SiO_2 layers.

With an efficient cleaning solution (HF 0.1%), the first 50 nm contaminated oxide can be removed, leaving a copper free surface oxide (Cu level under 10^{11} at/cm²). To optimize dielectric process deposition and obtain an etch stop layer free of Cu, we have to control the fluorine contamination of the equipment.

References

[1] Edelstein D, et al. Proc. IEEE IEDM 1997:773.
[2] Venkatesan S, et al. Proc. IEEE IEDM 1997:769.
[3] Marcadel C, et al. Proc. IEEE IEDM 1997:405.
[4] Labiadh A, et al. Proc E-MRS Symp J 1996;66:369.
[5] Tardif F, et al. Proc E-MRS Symp J 1996;66:195.

PERGAMON

Solid-State Electronics 43 (1999) 1019–1023

SOLID-STATE ELECTRONICS

The interaction of metals and barrier layers with fluorinated silicon oxides

Sarah E. Kim, Christoph Steinbrüchel*

Department of Materials Engineering and Center for Integrated Electronics and Electronic Manufacturing, Rensselaer Polytechnic Institute, Troy, NY 12180, USA

Received 19 July 1998; received in revised form 4 January 1999; accepted 16 January 1999

Abstract

Fluorinated silicon oxide (FSG) films with varying fluorine (F) content were prepared by plasma-enhanced chemical vapor deposition (PECVD) using TEOS, O_2, and either C_2F_6 or NF_3. Metal films (Al, Cu–1% Al, Cu) were deposited on the FSG either directly or with a barrier layer (Ta, TaN) between the metal and the FSG. In addition, undoped PECVD SiO_2 was studied as a capping layer on the FSG. Compositional depth profiles were obtained with both XPS and nuclear reaction analysis (NRA). F diffused rapidly through Ta, TaN and Al, and reacted with Al on the top surface. On the other hand, regardless of barrier layers, Cu showed almost no reaction with F. F diffusion was also observed through PECVD SiO_2. © 1999 Elsevier Science Ltd. All rights reserved.

1. Introduction

Among many new candidates for interlayer dielectrics (ILD), there has been an increased interest in low dielectric constant FSG films [1–10]. An obvious advantage of these films is their similarity to standard oxide films and hence their ability to be readily integrated into the fabrication process. However, it is acknowledged that FSG films can give rise to reliability problems due to the corrosive interaction of F from FSG with metals.

So far, only few studies have been reported on metal–fluorine reactions in metal/FSG systems. Passemard et al. [6], who studied Ti/TiN/AlCu/TiN/FSG structures, found that the corrosion affected all the metallic layers including AlCu when the F content was above 5 at%. Choi et al. [8] observed the corrosion of Ti or TiN in Al/Ti (or TiN)/FSG, but not in

the Al lines. Labiadh et al. [9] reported that in the absence of a barrier, neither Cu diffusion into FSG films nor F diffusion into Cu was detected. However, they found F diffusion into the barrier layers, i.e. into Ti and TiN.

The present paper is focused on the effect of F in the metal/dielectric system by studying interface reactions and diffusion of F into the metal (Al, Cu–1% Al, Cu). In order to reduce F diffusion various barrier layers (Ta, TaN) between the metal and the FSG, as well as a method of surface plasma treatment on FSG films [10], were also investigated.

2. Experimental details

The FSG films were deposited in a commercial single wafer parallel-plate PECVD reactor (Applied Materials Precision-5000 system) using TEOS, O_2, and either C_2F_6 or NF_3. Gas flow rates for TEOS (with He as a carrier gas) and oxygen were 500 and 250 sccm, respectively. Gas flow rates for C_2F_6 and NF_3 were

---

* Corresponding author. Fax: +1-518-2768761.
E-mail address: chris@unix.cie.rpi.edu (C. Steinbrüchel)

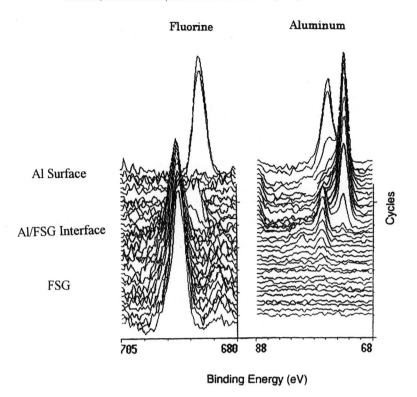

Fig. 1. XPS binding energy vs. depth into the Al/FSG/Si sample. Spectra are recorded in regular time intervals during sputter-etching of the sample.

varied from 0 to 100 sccm. The susceptor temperature was set to 390°C, deposition pressure to 3 Torr, and rf power to 250 W. The deposition process and the characteristics of the deposited FSG films have been described previously [10].

The F-metal (Al, Cu–1% Al, Cu) reaction was investigated by both XPS and NRA depth profiling after various heat treatments. NRA yields quantitative information on F content as well as low-resolution compositional depth profiles [11]. PECVD SiO_2 and sputtered Ta and TaN layers were examined as diffusion barriers for F between metal and the FSG. Films were analyzed as deposited and after annealing at 400°C for 5 h in N_2.

3. Results and discussion

In most cases, F displaces other halogens from their compounds, and it also displaces oxygen from most oxides, although Al_2O_3 is fairly stable towards attack by F. Thus, F can be expected to react strongly with Al when the two are in direct contact. However, we have found that there is no strong interface reaction. Rather, F diffuses readily through Al during annealing and reacts with Al and O at the top Al surface. NRA results show a F pile-up at the Al surface, but almost no F in the bulk of the films [10].

These findings are in accord with the XPS results in Fig. 1. It is clear that the Al binding state at the Al/FSG interface is different from the one at the surface of the Al film and also there is no F in the bulk of the Al films. The small peak at the interface corresponds to AlF_3, which has ~76.2 eV binding energy [12], while the peak at the surface with a ~75.5 eV binding energy does not correspond to AlF_3, but rather is between Al_2O_3 (~74.6 eV binding energy [12]) and AlF_3. Thus, at the interface F appears to react with the thin initial layer of Al as deposited, while at the top surface the F which has diffused through Al reacts with both Al and O from the native aluminum oxide.

In order to prevent or at least reduce F diffusion from FSG films, various plasma treatments of the FSG before metal deposition were investigated. The treatment consisted of a light etch with a CF_4/O_2 plasma under F-deficient conditions, and was designed

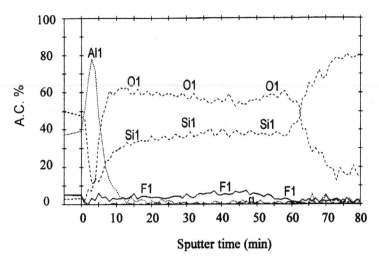

Fig. 2. XPS depth profile (Al/PTEOS SiO$_2$/FSG/thermal SiO$_2$/Si).

to deplete the FSG surface of F and thus to promote the metal–oxygen interaction. Optimal plasma conditions produced a significant reduction in F diffusion into Al without increasing the dielectric constant of the FSG [10].

Among possible barriers, a 300 Å PTEOS SiO$_2$ capping layer was first studied to prevent F diffusion into Al. As shown in Fig. 2, during annealing F diffused through the capping layer and reacted with Al at the surface. We note that Shapiro et al. [7] reported similar F diffusion even through 1 μm PTEOS SiO$_2$. With 300 Å of Ta between Cu–1% Al and FSG, F diffused

through the Ta and was observed in the bulk of Cu–1% Al rather than at the surface, as shown in Fig. 3. This suggests that F diffuses via Al located at Cu grain boundaries. Hence, some F exists in the bulk of Cu–1% Al.

Figs. 4 and 5 indicate that a 300 Å Ta layer produced significant improvement in controlling F diffusion into Al compared to a 300 Å TaN layer. However, a trace of F was still observed on the top of the Al even with a Ta barrier. The shoulder at a depth of 0.7 μm in Fig. 4 is evidence of some F accumulation and presumably reaction of F with the TaN. There

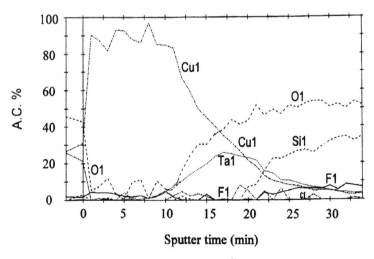

Fig. 3. XPS depth profile (FSG/~300 Å Ta/Cu–1% Al).

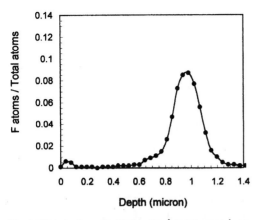

Fig. 4. NRA depth profile (FSG/~300 Å TaN/~5000 Å Al).

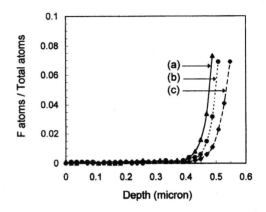

Fig. 6. NRA depth profile (a) plasma treated FSG/~300 Å Ta/~4200 Å Cu, (b) FSG/~300 Å Ta/~4200 Å Cu, (c) FSG/~300 Å TaN/~4200 Å Cu.

was no significant difference between non-treated films and plasma treated films with a Ta barrier layer. Thus we conclude that Ta has better barrier properties, but neither Ta nor TaN can be a reliable barrier for F diffusion into Al. Preliminary results with a 300 Å TiN barrier indicate no fluorine diffusion into Al, but TiN itself also reacts with fluorine at the Al/TiN interface [13].

Fig. 6 demonstrates that there was almost no F diffusion through Cu and at most a very small amount of F was present in the bulk of the Cu. There was no difference between Ta (Fig. 6(b)) and TaN (Fig. 6(c)) as a barrier, nor between non-treated films (Fig. 6(b)) and plasma treated films (Fig. 6(a)). Therefore, in comparison to Al, one may conclude that the reason for no reaction between F and Cu is not the barrier, but rather the lack of reactivity of the copper itself.

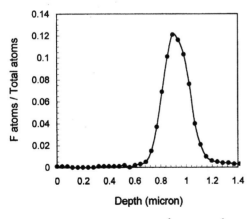

Fig. 5. NRA depth profile (FSG/~300 Å Ta/~5000 Å Al).

4. Summary

FSG films deposited by PECVD were investigated for interaction with metals, namely Al, Cu–1% Al, and Cu. For Al and Cu–1% Al, diffusion of F atoms through the metal film is rapid at typical annealing conditions and is noticeable even at room temperature over a period of weeks. At the same time, no major reaction occurs at the metal–dielectric interface. With Al, the diffused F accumulates on the top surface of the metal film and no F is present in the bulk of the film. With Cu–1% Al, no surface enrichment but a larger bulk concentration of F are observed. With Cu, almost no F was observed both in the bulk and on the surface of Cu. A Ta barrier produced a significant reduction of F diffusion into metal compared to TaN. However, a trace of F was still observed on top of the Al even with a Ta barrier. TiN did not show fluorine diffusion through Al compared to TaN. Both TaN and TiN reacted with F. A suitable plasma treatment of the FSG before metal deposition inhibits substantially the diffusion of F into the metal without increasing the dielectric constant of the FSG. F also diffuses easily through undoped PECVD oxide.

Acknowledgements

The authors acknowledge A. Kumar and Dr. H. Bakru (SUNY at Albany) for their NRA measurements. This work was supported in part by the Semiconductor Research Corporation (SEMATECH Center for Advanced Interconnect Science and Technology at RPI) and by Texas Instruments.

References

[1] Usami T, Shimokawa K, Yoshimaru M. Jpn J Appl Phys Part 1 1994;33:408.

[2] Song J, Ajmera PK, Lee GS. Appl Phys Lett 1996;69:1876.

[3] Nakasaki Y, Miyajima H, Katsumata R, Hayasaka N. Jpn J Appl Phys Part 1 1997;36:2533.

[4] Yoo WS, Swope R, Mordo D. Jpn J Appl Phys Part 1 1997;36:267.

[5] Chang KM, Wang SW, Li CH, Yeh TH, Luo JJ. J Electrochem Soc 1997;144:1754.

[6] Passemard G, Fugier P, Noel P, Pires F, Demolliens O. Microelectronic Engineering 1997;33:335.

[7] Shapiro MJ, Matsuda T, Nguyen SV, Parks C, Dziobkowski C. J Electrochem Soc 1996;143(7):L156.

[8] Choi JH, Hwang BK, Shin HJ, Chung UI, Lee SI, Lee MY. Proc IEEE VMIC Conf 1995;397.

[9] Labiadh A, Braud F, Torres J, Palleau J, Passemard G, Pires F, Dupuy JC, Dubois C, Gautier B. Microelectronic Engineering 1997;33:369.

[10] Kim SE, Steinbrüchel C, Kumar A, Bakhru H. Mater Res Soc Sym Proc 1998;511, 191.

[11] Kumar A, Bakhru H, Haberl AW, Carpio RA, Ricci A. 14th Int. Conf. on the Application of Accelerators in Research and Industries, 1996.

[12] Handbook of X-ray Photoelectron Spectroscopy, Physical electronics division. Perkin-Elmer Co, 1992.

[13] Kim SE, Steinbrüchel C, Kumar A, Bakhru H. Proc SPIE Conf on Multilevel Interconnect Technology II 1998;3508:51.

REFERENCES

PERGAMON

Solid-State Electronics 43 (1999) 1025–1029

SOLID-STATE ELECTRONICS

Evaluation and localization of oxygen in thin TiN layers obtained by RTLPCVD from TiCl$_4$–NH$_3$–H$_2$

L. Imhoff*, A. Bouteville, H. de Baynast, J.C. Remy

Laboratoire de Physico-Chimie des Surfaces, ENSAM-CER d'Angers, 2 Bd du Ronceray, BP 3525, 49035 Angers, France

Received 16 July 1998; received in revised form 16 November 1998; accepted 21 December 1998

Abstract

Oxygen is well known to play a significant role in the barrier behavior of titanium nitride layers for the IC metallization. TiN thin films are deposited by rapid thermal low pressure chemical vapor deposition from the TiCl$_4$–NH$_3$–H$_2$ gaseous phase onto silicon substrate. Oxygen contamination level lies in the range of 3.5–10% when the deposition temperature decreases from 800–500°C. The aim of this paper is to localize oxygen in the layer by a variety of complementary analytical techniques: Rutherford backscattering spectrometry, nuclear reactive analysis, Auger electron spectroscopy, X-ray photoelectron spectroscopy, X-ray diffraction and scanning electron microscopy. These analyses show that O is diffused in the film from the surface, and for low temperature films, O is most present at the interface TiN/Si. With small O contents (< 3%), this contamination is situated at grain boundaries, and for higher contents, O is diffused in the grains. For high temperatures, a columnar structure is observed, and for low temperatures, a granular structure is observed. The layers are covered by copper and work well as a diffusion barrier between copper and silicon. © 1999 Published by Elsevier Science Ltd. All rights reserved.

1. Introduction

Titanium nitride possesses a unique combination of properties making it an attractive choice for various thin film applications [1]. In particular, in the microelectronics industry, TiN has been shown to be a very useful diffusion barrier layer between metal and silicon [2]. Ti is an oxygen getter, and during the deposition of the film or after air exposure, TiN tends to incorporate oxygen into the film [3]. The presence of oxygen, incorporated in TiN, has been reported to enhance the barrier properties of TiN layers [4]. The aim of this paper is to localize oxygen in layers obtained on silicon substrate by rapid thermal low pressure chemical vapor deposition (RTLPCVD): on the surface, at the interlayer TiN/Si, in the grains or at the grain boundaries.

2. Experiment

The deposition apparatus is a modified commercial cold wall RTLPCVD JIPELEC machine described elsewhere [5]. Ammonia, titanium tetrachloride and hydrogen are used in the gaseous phase. NH$_3$ is introduced around the substrate quite near the surface, while TiCl$_4$ is introduced with H$_2$ 10 cm above the substrate middle. The substrate is a 100-mm diameter ⟨100⟩ oriented silicon wafer. Deposition temperature and total pressure lie in the range of 500–800°C and 200–500 mTorr (27–70 Pa), respectively.

* Corresponding author. Lab. de Recherches sur la Réactivité des Solides, 9 avenue Alan Savary, BP 47870, 21078 Dijon Cedex, France. Tel.: + 33-3-8039-6161; fax: + 33-3-8039-6132.

E-mail address: luc.imhoff@u-bourgogne.fr (L. Imhoff)

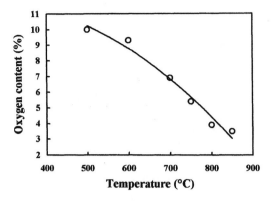

Fig. 1. Variation of oxygen content in relation to the temperature.

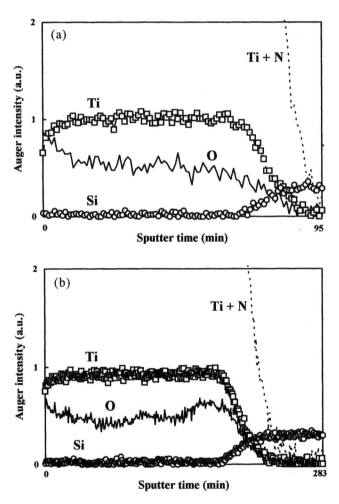

Fig. 2. AES depth profiles for layers obtained with deposition temperatures of (a) 800°C and (b) 500°C.

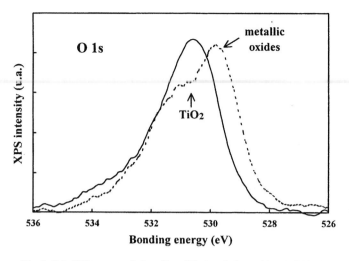

Fig. 3. O 1s XPS spectrum before (dotted line) and after etching (solid line).

The films were characterized by a variety of complementary analytical techniques: Rutherford backscattering spectrometry (RBS, 2.4 Mev He^{2+} ion beam), nuclear reactive analysis (NRA), Auger electron spectroscopy (AES, 5 kV electron beam) combined with a 3 keV Ar^+ ion sputtering beam, X-ray photoelectron spectroscopy (XPS, 10 kV Mg anode X-ray source), X-ray diffraction (XRD, Cu Kα beam) and scanning electron microscopy (SEM).

3. Results and discussions

From XPS and RBS analyses, chlorine contamination can be evaluated to be less than 0.1%. This good result is attributed to the separate introduction of $TiCl_4$ [5]. From RBS and NRA analyses, oxygen contamination level increases from 3.5% at 800°C to 10% at 500°C (Fig. 1). All the RBS analyses reveal a Ti/N ratio very close to 1 as can be expected for chemical vapor deposits [6]. Nevertheless, the accuracy of the RBS results is not greater than 3%.

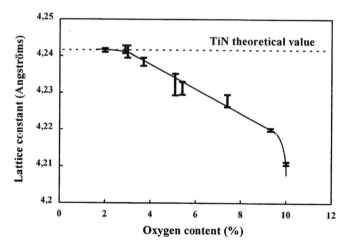

Fig. 4. Evolution of the lattice constant in relation to oxygen content.

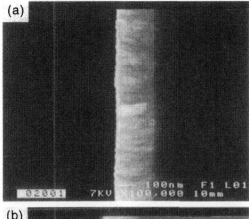

Fig. 5. SEM pictures of (a) the columnar structure and (b) the granular structure of the layers.

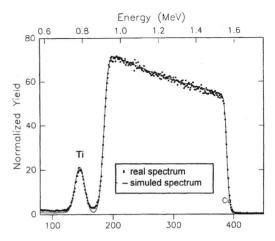

Fig. 6. RBS spectra of a Cu/TiN/Si structure.

In Fig. 2, AES depth profiles are represented for deposition temperatures 800°C (Fig. 2(a)) and 500°C (Fig. 2(b)). At 800°C, the O signal monotonously decreases with depth, which can signify that the O has been diffused from the surface [7]. For temperatures inferior to 600°C (Fig. 2(b)), the O signal decreases with depth in the first half part of the layer and then shows an enrichment of O at the interlayer TiN/Si. So, for the first part of the layer, the O decrease is identically interpreted to the decrease observed at 800°C. The enrichment of O at the interlayer can be explained by the H_2 cleaning carried out just before the deposition of the layers. This cleaning is made at the same temperature of the deposition by injecting H_2 on the surface of the wafer, and at 500°C, the cleaning is less efficient than at 800°C.

Fig. 3 shows the O 1 s XPS spectrum of a layer before and after a 1 minute etching with a 2 KeV Ar^+ ion sputtering beam. Before etching, the spectrum is composed of two peaks. The first is situated at 530.6

eV and indicated TiO_2 type bonding. The second is situated between 529.5 and 530 eV and can originate from the adsorption of various metallic oxides [8]. After etching, only the first peak is still present. The difference between the two curves shows that there is much more O present at the surface than in the layer.

The Fig. 4 shows the evolution of the lattice constant given by XRD in relation to the oxygen content given by NRA. When the quantity of O contained in the layer is smaller than 3%, the lattice constant is equal to the theoretical value of TiN. When this quantity of O is higher than 3%, the lattice constant decreases. The diminution of the lattice constant can be attributed to the substitution of nitrogen to oxygen in the TiN lattice [9], because whatever the type bonding may be, atomic radius of O is always smaller than atomic radius of N [10]. Thus, for O quantities smaller than 3%, O is probably situated at the grain boundaries, and for quantities larger than 3%, O is diffused in the grains.

The Fig. 5 shows SEM pictures of two layers. The first (Fig. 5(a)) is obtained at 800°C and presents a columnar structure, with a mean diameter of columns of 40 nm. The second (Fig. 5(b)) is obtained at 500°C and presents a granular structure, with a mean diameter of grains of 30 nm. The second structure is theoretically a better barrier, because the diffusion paths are reduced with a granular structure in relation to columnar structure [11]. The O quantities can be linked to the structure because, when the temperature decreases, the O quantity increases and, at the same time, the structure changes from columnar to granular, so the grain boundaries increase.

In order to check the barrier properties of the layers, the samples have been covered by 520 nm of copper, annealed by RTP at 500°C for 1 min, and then analyzed by RBS. Fig. 6 shows the RBS spectrum of such a sample, and a simulated spectrum obtained on the assumption that there is no interdiffusion between copper and the TiN layer. The two curves are similar, so this structure is found to be stable up to 500°C. Moreover the resistivity of the TiN layers obtained at 500°C is smaller than 250 $\mu\Omega$ cm, which is a good result, considering the small thickness of the TiN layers used like barrier diffusion.

4. Conclusion

TiN films are chemically vapor-deposited from the $TiCl_4$–NH_3–H_2 gaseous phase onto silicon substrate. Oxygen contained in the layers increases from 3.5% at 800°C to 10% at 500°C. The analyses show that O diffuses in the film from the surface, and for films obtained at low temperature, O is most present at the interface TiN/Si. When the O content is smaller than 3%, O is situated in the grain boundaries, whereas O content is larger than 3%, O is diffused in the grains. Finally, a structure modification is observed when the deposition temperature decreases: the structure changes from columnar to granular.

Acknowledgements

The authors gratefully thank J. Watson for reviewing the translation of this article, J. Perriere for the RBS and NRA analyses (Groupe de Physique des Solides des Universités Paris 6 et 7), B. Grolleau for the surface analyses and especially C. Cardinaud for AES and A. Barrault for SEM (Institut des Matériaux de Nantes).

References

[1] Kurtz SR, Gordon RG. Thin Solid Films 1986;140:277.
[2] Sherman A. Jpn J Appl Phys 1991;30(12B):3553.
[3] Olowolafe JO, Li J, Mayer JW, Colgan EG. Appl Phys Lett 1991;58:469.
[4] Lu JP, Hsu WY, Hong QZ, Dixit GA, Luttmer JD, Havemann RH, Magel LK. J Electrochem Soc 1996;143(12):L279.
[5] Imhoff L, Bouteville A, Remy JC. J Electrochem Soc 1998;145(5):1672.
[6] Fouilland L, Imhoff L, Bouteville A, Benayoun S, Remy JC, Perrière J, Morcrette M. Surf Coat Technol 1998;100–101:146.
[7] Eizenberg M, Littau K, Ghanayem S, Liao M, Mosely R, Sinha AK. J Vac Sci Technol 1995;13(3):590.
[8] Moulder JF et al. In: Chastin J, editor. Handbook of X-ray photoelectron spectroscopy. Perkin–Elmer Corporation, 1990.
[9] Makino Y, Nose M, Tanaka T, Misawa M, Tanimoto A, Nakai T, Kato K, Nogi K. Surf Coat Technol 1998;98:934.
[10] Kittel C. In: Dunod Université, editor. Physique de l'état solide. 5th ed., 1983.
[11] Nicolet M-A. , Thin Solid Films 1978;52:415.

PERGAMON

Solid-State Electronics 43 (1999) 1031–1037

SOLID-STATE
ELECTRONICS

Observation of current polarity effect in stressing as-formed sub-micron Al–Si–Cu/TiW/TiSi$_2$ contacts

Li-Zen Chen, Klaus Y-J. Hsu*

National Tsing Hua University, Department of Electrical Engineering, Hsinchu, Taiwan

Received 27 July 1998; received in revised form 14 December 1998; accepted 21 January 1999

Abstract

Formation of good silicide contacts becomes more important but difficult as the contact size continues shrinking toward the deep sub-micron regime. At the same time, higher current density, which may easily appear in small regions, could pose strong impact to the long-term reliability of sub-micron contacts. In this work, high current density stress experiments were conducted on the Al–Si–Cu/TiW/TiSi$_2$ contacts with the size ranging from 0.5×0.5 μm^2 down to 0.25×0.25 μm^2. The self-aligned silicide contacts were formed by using collimated sputtering, E-beam lithography, RTA, and RIE techniques. The silicide contacts were sintered at 400°C for 30 min. Cross-bridge Kelvin resistor structures were formed for electrical stressing and contact resistance measurement. One-way and two-way stressings were performed at high current density ($\sim 10^7$ A/cm^2) and the contact resistance was measured periodically at low current density during the stressing to monitor the evolution. It was found that the initial resistance of as-formed contacts was higher than expected. This is probably due to the difficulty of forming good interfaces in the small contact region by sputtering and that the sintering temperature may not be high enough to smear out the imperfection. The stressing was found to anneal the contacts. With electrons flowing from metal layer into the contact window, the contact resistance was reduced more efficiently than with reverse current of the same density. Stressed first by reverse current then by normal current, the resistance showed a two-step reduction with a significant transition at the switch of current polarity. For prolonged stressing, the contacts were gradually degraded and the reverse current induced more severe damage. These observations indicate strong electromigration effect at the small contacts. © 1999 Published by Elsevier Science Ltd. All rights reserved.

1. Introduction

As the integration level of microelectronic integrated circuits gets higher and higher, the feature size of interconnects and contacts becomes smaller. Also, multiple levels of metal wiring become necessary. Consequently, the number of small metal-semiconductor and metal–metal interfaces grows enormously. In the mean time, the current density flowing through these interfaces can easily reach a high value beyond, for example, 1×10^5 A/cm^2. Under this high current density condition, it is unavoidable that one has to seriously consider the impact of electromigration on the reliability of the interfaces.

In the regime of ultra-large scale of integration (ULSI), another issue related to the small interfaces requires attention. Since the size of the contacts in the transistors shrinks down to the deep sub-micron level, conventional metal contact resistance may become large and degrade the device performance. It is well known that smaller contact resistance may be obtained

* Corresponding author. Tel.: +886-3-5742594; fax: +886-3-5715971.
E-mail address: yjhsu@ee.nthu.edu.tw (K.Y-J. Hsu)

0038-1101/99/$ - see front matter © 1999 Published by Elsevier Science Ltd. All rights reserved.
PII: S 0 0 3 8 - 1 1 0 1 (9 9) 0 0 0 2 0 - 9

by replacing the conventional metal contacts by silicide contacts. The potential of obtaining small contact resistance by using silicide has been demonstrated. Specific contact resistance, defined as the contact resistance multiplied by the contact area, down to 1×10^{-8} Ω cm^2 was obtained by using Ti silicidation and Ta silicidation [1–2]. However, for deep sub-micron contacts, it is often difficult to form good silicide. Firstly, the thin film coating in small contact windows becomes a challenging task. Not only conformity is not easy to achieve, but also the interfaces become less ideal. Secondly, the stress effect in the small contact windows may result in incomplete phase transition during the annealing process of silicidation, which leads to higher resistivity.

The present paper reports the results of electrically stressing sub-micron Ti silicide contacts, which is related to the two issues described above. It will be shown that at the initial stage of high-current stressing, contact resistance could be reduced if the contact was originally not perfectly formed. But long-term stressing degraded the contact again. In both cases, current polarity effect was observed, which indicates that electromigration plays an important role in the small silicide contacts.

2. Experimental

D-type Kelvin contact resistance test pattern was fabricated for the electrical stressing and the contact resistance measurement. Fig. 1 shows the schematic illustration of the test pattern. Note that among the three contact windows, only the smaller one is concerned because the current density it experiences is much higher than that experienced by the other two contacts. The width of the diffusion region was designed to match the contact window size so as to reduce the artifact in measuring the resistance that might be introduced due to lateral current spreading effect [3–4].

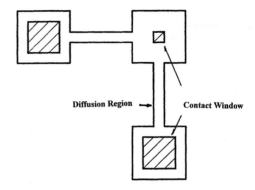

Fig. 1. Schematic illustration of the D-type Kelvin bridge structure.

To fabricate the test structure, 'mix and match lithography' technique was utilized. That is, we used electron beam lithography for defining the diffusion region and the contact windows, and G-line stepper was used for metal level pattering. N-type, 6″, (100)-oriented Si wafers with $\rho = 2 \sim 3$ Ω-cm were used as the substrates. After the standard RCA cleaning, the alignment mark for G-line stepper was placed on the wafer by using reactive ion etching (RIE). Then a 450 nm thick field oxide layer was grown by wet oxidation. After placing the alignment mark for electron beam lithography, the diffusion region was opened by the electron beam lithography and the RIE on field oxide. Then a thin dry oxide (the pad oxide) of 25 nm in thickness was grown on top of the diffusion region. BF$_2^+$ ions were implanted through the pad oxide at an incident energy of 40 KeV and a dose of 5×10^{15} cm^{-2}. Thermal annealing at 800°C for 30 min followed by a rapid thermal annealing (RTA) at 1050°C for 20 s was applied to activate the implanted ions in the diffusion region. After removing the pad oxide, a new oxide layer with a thickness of 200 nm was deposited on the diffusion region by using low-temperature plasma-

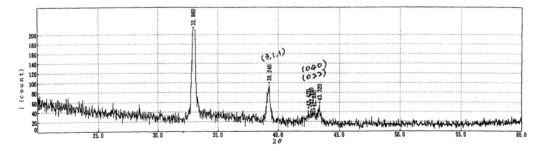

Fig. 2. X-ray diffraction pattern of the Ti silicide after the two-step RTA.

enhanced chemical vapor deposition (PECVD). This oxide layer is where the contact windows were placed and it was designed to be relatively thick in order to prevent any possible leakage current path. PECVD was chosen in order to avoid disturbing the doping in the diffusion region. The contact windows were subsequently opened through the deposited oxide layer by using electron beam lithography and RIE, and additional ion implantation and annealing were done for the contact windows to increase the doping concentration at the contacts. A 30 nm Ti film and a 30 nm TiN films were then consecutively deposited by collimated sputtering. During the sputtering, the substrate temperature was held at 400°C. The collimator was used because it is beneficial to coating thin films into the holes with high aspect ratios. The TiN film was deposited to protect the Ti film from being consumed during the subsequent two-step RTA process in N_2 ambient. The two-step RTA was applied to form Ti silicide in the contact windows. The first RTA was done at 650°C for 30 s to form the C49-phase Ti silicide. Solution of $H_2SO_4/H_2O_2 = 3 : 1$ was applied before the second RTA step to remove the TiN film and the unreacted Ti on the oxide. The second RTA was done at 850°C for 30 s to transform the C49-phase Ti silicide into C54 phase. After the formation of Ti silicide in the contact windows, a TiW barrier layer of 100 nm in thickness was deposited by sputtering, followed by the sputtering of a 500 nm thick Al–Si–Cu (98.5–1–0.5%) metal layer and a 30 nm thick

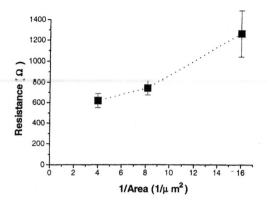

Fig. 3. Contact resistance values of as-formed Ti silicide contacts.

TiN protection layer. Finally, the metal layers were patterned and then sintered at 400°C for 30 min.

The samples were subjected to several characterizations such as sheet resistance measurement, α-step measurement, SIMS profiling, SEM investigation, X-ray diffraction, contact resistance measurement, and electrical stress test. In the electrical stressing, the current direction when electrons flow from metal layers into the diffusion region is defined as positive, and the reverse direction is negative. One-way and two-way stressings were performed at high current density ($\sim 10^7$

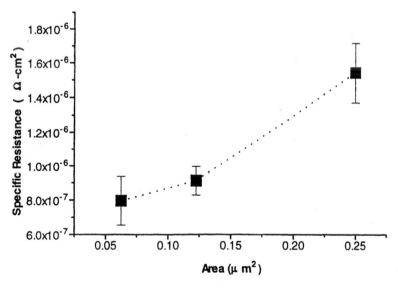

Fig. 4. Specific contact resistance values of as-formed Ti silicide contacts.

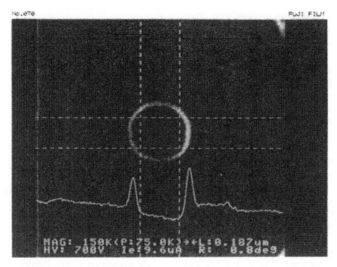

Fig. 5. SEM image of the 0.25 × 0.25 μm contact window after the RIE process. The distance between the two vertical dashed lines is 0.187 μm.

A/cm²) and the contact resistance was measured periodically at low current density ($< 10^4$ A/cm²) during the stressing to monitor the evolution. The stressing was conducted at room temperature.

3. Results and discussion

By the correlation between α-step measurement, SIMS measurement, and TSuprem4 simulation, the thickness of TiSi₂ was estimated to be 40 nm. The boron concentration in the silicide region was found to reach 1×10^{21} cm⁻³. Fig. 2 shows the X-ray diffraction pattern of the Ti silicide after the two-step RTA. It is clear that the characteristic peaks corresponding to the C54(311), (022), and (040) orientations are present, which indicates the formation of C54 phase. At the same time, four-point probing on the Ti silicide films gave a sheet resistance value of 36.5 Ω/□ after the first RTA. After the second RTA, the value was reduced to 11.5 Ω/□ which, combining the 40 nm thickness, gives a resistivity of 46 μΩ-cm. Although these results indicate the presence of C54 TiSi₂, the resistivity value is higher than those reported in Refs [1,2]. This is probably due to that our silicide was deposited after implantation. Therefore, there was no implantation-through-metal (ITM) process occurred in our samples. Being lack of the inter-mixing effect by ITM, the silicidation by RTA might be less effective and the resultant resistivity might be higher. Our previous experience with the ITM process showed that the resistivity of C54 Ti silicide can be around 15 μΩ-cm [5].

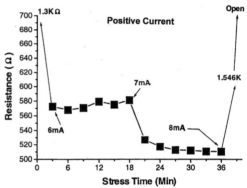

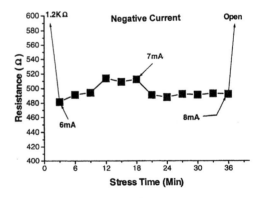

Fig. 6. Stress test result for 0.25 × 0.25 μm contacts. The current magnitude is changed after the total stressing time reaches 18 and 36 min.

Fig. 3 shows the relation between the measured contact resistance of as-formed contacts and the designed contact area. Fig. 4 shows the corresponding specific contact resistance values. From these Figures, it seems that the linear Ohm's law does not hold and the resistance is too large. As mentioned above, it is difficult to perform good sputtering in the small contact windows with high aspect ratios. Therefore, the interfaces in the multiple metal layers might not be in a perfect condition even after the sintering process. Besides, SEM investigation showed that the actual contact window area may be smaller than the designed value. Fig. 5 shows the image of the 0.25×0.25 μm contact window after the RIE process. Plus the fact that the resistivity was higher in our as-formed silicide, the high value and non-linearity of the measured resistance should not be surprising.

When the as-formed contacts were subjected to high current stressing, interesting polarity effect was

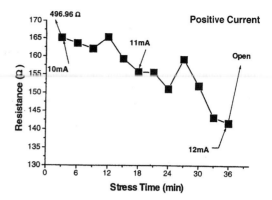

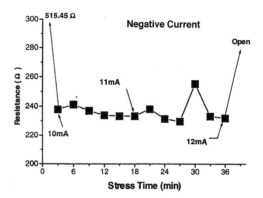

Fig. 8. Stress test result for 0.5×0.5 μm contacts.

observed. Fig. 6 shows a typical stress test result for 0.25×0.25 μm contacts. In the Figure, the initial stress current was set as 6 mA ($\sim 10^7$ A/cm^2). The stress period between every two consecutive sampling points was set as 3 min. For both current directions, the as-formed contacts experienced drastic reduction of contact resistance values at the beginning of the high-current stressing. Then, after the stress current was switched to 7 mA, another resistance reduction was observed. But this reduction is larger for positive current than for negative current. When the stress current was further increased to 8 mA, the contacts were burnt out for both directions. It seemed that during the stress test, there were two driving forces determining the behavior of contact resistance at the same time. One was the Joule heating effect that is isotropic. The other was the electromigration effect that may be directional. In order to reduce the isotropic heating effect, another stress test on the contacts of the same size was conducted at reduced current density. Fig. 7 shows the result. The difference between the two current polarities is obvious. Positive current is more effective in

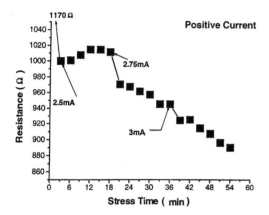

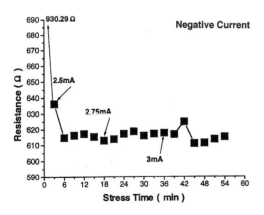

Fig. 7. Stress test result for 0.25×0.25 μm contacts at reduced current density.

reducing the contact resistance value of the as-formed Ti silicide contacts. This may be attributed to that in addition to the healing effect of Joule heating, the electrons flowing from metals into Si (positive stress current) enhance the intermixing of Ti and Si, thus further improving the silicidation and reducing the resistance. The same phenomenon was observed for larger contacts. Fig. 8 is the result for the $0.5 \times 0.5 \ \mu m^2$ contacts. The polarity effect due to electromigration can be further illustrated by the result shown in Fig. 9. In this Figure, the sampling period was reduced from 3 min down to 0.5 s to reduce Joule heating, and the stress current polarity was reversed after 30 s of stressing. The characteristics of polarity effect observed in Figs. 7 and 8 are combined in this Figure. It is clear that the negative current did almost nothing in healing the as-formed contacts, while the positive current reduced the contact resistance quite efficiently.

Although high-current stress, especially with positive current, was shown to be able to improve the quality

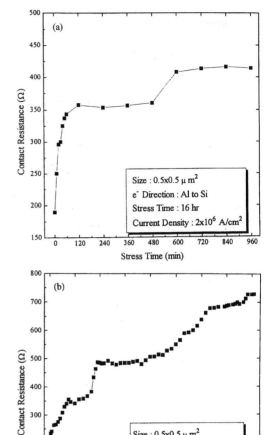

Fig. 10. Prolonged high-current stress result for the healed $0.5 \times 0.5 \ \mu m^2$ contacts: (a) positive current; (b) negative current.

of the Ti silicide contacts, it should be noted that long-term stress still degrades the contacts. And the degradation also exhibits polarity effect. Fig. 10 shows the prolonged high-current stress result for the healed $0.5 \times 0.5 \ \mu m^2$ contacts. It can be observed that while positive stress current tends to saturate the degradation, negative current continually destroys the stability of the contacts. With negative stress current, silicon atoms could be pushed into the $TiSi_2$–TiW–metal multi-layers and create surface pits or accumulate somewhere in the contact. This might be one of the causes that contribute to the degradation. It should be

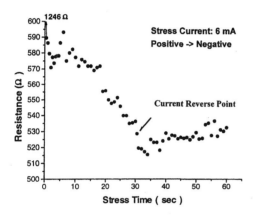

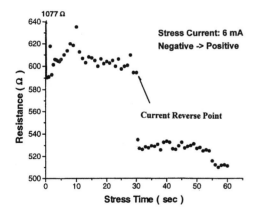

Fig. 9. Two-way stress test result for $0.25 \times 0.25 \ \mu m$ contacts.

noted that the polarity effect of contact reliability was also observed in other silicide contacts. Huang et al. [6] observed the effect in Ni_2Si contacts even for low current-density (5×10^4 A/cm^2) stress.

4. Conclusion

High-current stress was performed on as-formed Ti silicide contacts. It was found that short-term stress could reduce the contact resistance but long-term stress degraded the contacts again. Clear current polarity effect was observed in both cases. Positive stress current is more effective in healing the as-formed contacts during the short-term stress. But negative stress current is more harmful to the contacts during the long-term stress. This indicates that electromigration plays an important role in the reliability of small silicide contacts.

References

[1] Kotaki H, Mitsuhashi K, Takagi J, Akagi Y, Koba M. Low resistance and thermally stable Ti-silicide shallow junction formed by advanced 2-step rapid thermal processing and its application to deep submicron contact. Jpn J Appl Phys, Part 1 1993;32:389–95.

[2] Yamada K, Tomita K, Ohmi T. Formation of metal silicide-silicon contact with ultralow contact resistance by silicon-capping silicidation technique. Appl Phys Lett 1994;64:3449–51.

[3] Loh WM, Saraswat K, Dutton RW. Analysis and scaling of Kelvin resistors for extraction of specific contact resistivity. IEEE Electron Device Lett 1985;EDL-6:105–8.

[4] Loh WM, Swirhun SE, Crabbé E, Saraswat K, Swanson RM. Accurate method to extract specific contact resisitivity using cross-bridge Kelvin resistors. IEEE Electron Device Lett 1985;EDL-6:441–3.

[5] Lin CC, Chen WS, Huang HL, Hsu KYJ, Liou HK, Tu KN. Reliability study of sub-micron titanium silicide contacts. Appl Surf Sci 1996;92:660–4.

[6] Huang JS, Liou HK, Tu KN. Polarity effect of electromigration in Ni_2Si contacts on Si. Phys Rev Lett 1996;76:2346.

PERGAMON

Solid-State Electronics 43 (1999) 1039–1044

SOLID-STATE
ELECTRONICS

Cobalt silicide thermal stability: from blanket thin film to submicrometer lines

A. Alberti[b,*], F. La Via[a], M.G. Grimaldi[b], S. Ravesi[c]

[a]*CNR-IMETEM, Stradale Primosole 50, 95131, Catania, Italy*
[b]*Physics Department, Catania University, Corso Italia 57, Catania, Italy*
[c]*S-T Microelectronics, Stradale Primosole 50, 95131, Catania, Italy*

Received 8 July 1998; accepted 19 September 1998

Abstract

The effect of lateral dimension scaling on the thermal stability of $CoSi_2$ layers reacted on pre-amorphized chemical vapour deposited silicon has been demonstrated. Resistance measurements on both blanket and patterned silicide 100 nm thick have been performed in the temperature range between 850 and 1100°C. The annealing effect on the silicide linewidth and on the grain size distribution was studied by transmission electron microscopy plan-view. A strong correlation between line resistance and morphological change during the high temperature annealing has been found. The thermal degradation activation energies in blanket and patterned $CoSi_2$ films have been compared and explained. Moreover, thermal stability dependence on line width on (001) silicon has been also proved and compared to previous results on polycrystalline substrate. © 1999 Published by Elsevier Science Ltd. All rights reserved.

1. Introduction

Silicides on polycrystalline Si (poly-Si) are widely used as low resistance gate electrodes and local interconnection. Particularly, cobalt disilicide has attracted a lot of attention in submicron silicon technology, because of its low resistivity (10–18 μΩ cm) and its excellent chemical stability compared to other silicides such as $TiSi_2$ [1]. However, the application of these compounds is limited by their finite thermal stability, which produces a detrimental effect on device characteristics. In fact, thermal degradation, consisting in the agglomeration of silicide film, leads to an increase in the silicide line sheet resistance [2,3,4] and, sometimes, to a degradation of the gate oxide integrity [5].

It is well known that degradation depends on many physical parameters and experimental conditions [6]. In particular, $CoSi_2$ thin film shows a worse thermal stability on polysilicon [7] than on amorphous Si and the latter is also more stable than on single crystal substrate [8]. Moreover, the dopant species implanted in the polysilicon layer seems to have a great influence on this property [9,10].

Only a few investigations have been performed on the thermal stability of silicides in submicrometer lines [11]. In this work we directly compare the thermal stability of patterned and blanket $CoSi_2$ films reacted on amorphous silicon. The difference in the deterioration process is investigated and the correlation between the increase of the silicide resistance and the change in the morphology of the layer is discussed.

---

* Corresponding author. Tel.: +39-095-59-1912; fax: +39-095-713-9154.
E-mail address: lavia@imetem.ct.cnr.it (A. Alberti)

0038-1101/99/$ - see front matter © 1999 Published by Elsevier Science Ltd. All rights reserved.
PII: S0038-1101(99)00021-0

Moreover, a lateral scaling effect on thermal stability for cobalt silicide reacted on (001) silicon is reported.

2. Experimental

2.1. Blanket silicide

On (001) silicon wafers a 100 nm SiO_2 layer was grown by dry oxidation. Then, 200 nm amorphous Si layer was chemical vapor deposited at 500°C. Prior to cobalt deposition, the substrate was further [4] amorphized with Si (50 keV) or BF_2 (90 keV) implant at a dose of 5×10^{15} ions/cm^2. After a HF dip, to remove the native oxide, 30 nm cobalt layer was deposited in a ultra high vacuum e-gun system with a base pressure lower than 2×10^{-9} Torr. The samples were annealed at 540°C for 30 s in forming gas (N_2:H_2, 90:10), etched in HNO_3 (4%) to remove the unreacted Co and finally annealed at 800°C for 40 s in N_2 flux. The resulting silicide thickness was 100 nm wide.

2.2. Strip on polysilicon

On a CVD amorphous silicon deposited with the previously decribed process, a thin Si_3N_4 layer was deposited by LPCVD. The samples were patterned and subject to a LOCOS process at 920°C for ~2 h to realize test patterns for electrical and structural characterization up to 1 μm wide. Prior to cobalt deposition, the substrate was implanted with 5×10^{15} cm^{-2} Si ions (50 keV) to completely amorfize the substrate. The two step reaction already described, produce a silicide layer 100 nm thick.

2.3. Strip on (100) silicon

Onto (001) silicon wafers a 100 nm oxide layer was CVD deposited. Test patterns for both electrical and structural characterization were opened up to 0.4 μm. Then, cobalt deposition and silicide reaction in two steps were performed, as previously described.

3. Results and discussion

3.1. Electrical measurements

The silicide resistance measurements were performed at room temperature on both blanket and patterned silicide after high temperatures anneal. The process window probed covers the temperature range between 850 and 1100°C. For each sample, the silicide resistance has been referred to that of the as grown sample (R_0) in both blanket and patterned silicide. Particularly, sheet resistance for blanket as reacted sili-

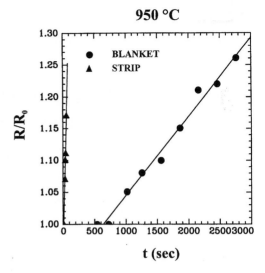

Fig. 1. Normalized resistance increase versus annealing time at 950°C. The deterioration rate in the blanket silicide is lower than in 1 μm wide strip.

cide is equal to 1.8–2 Ω/□, while 2400 Ω is the average resistance for patterned as grown silicide. Taking into account the line dimension and the silicide thickness, a silicide resistivity of about 20 μ Ω cm can be estimated. No difference between as reacted blanket silicide and line has to be expected because of lateral scaling effect

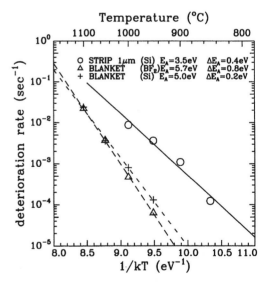

Fig. 2. Arrhenius plot of the deterioration rate. Different activation energies for blanket and patterned silicide were found.

(a)

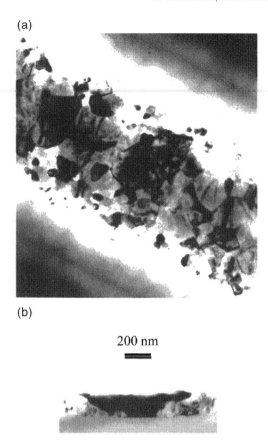

(b)

200 nm

Fig. 3. (a) TEM plan-view of the strip after anneal at 900°C for 200 s. An evident lateral roughness together with a mass loss at the strip edge were observed. (b) TEM cross section of the same sample shown in (a). A strong correlation between columnar silicide and underlying silicon grains in the as reacted sample was found. Upon anneal not only a silicide grain growth close to the strip center, but also an evident silicon grain diameter increase was observed.

on silicide resistivity [12]. Nevertheless, their behavior upon high temperature annealing is totally different. In both kind of samples the silicide resistance increases linearly together with the annealing time. But to achieve comparable resistance increase longer anneals in blanket layer than in stripes were necessary. In Fig. 1, a typical trend is reported. Particularly, for the blanket silicide 25 min anneal at 950°C is required in order to have a ∼10% resistance increase respect to the starting value, unlike the patterned silicide for which less than one minute is necessary. This means the blanket layer is more stable than the patterned one.

Because of the strong correlation found between structural evolution and electrical behavior of the sili-

cide layer upon annealing [13], we describe the silicide deterioration process in terms of resistance increase and the slope of the data linear fit represents the deterioration rate of the silicide layer [13].

The silicide thin film degradation is thermally activated, so the information about the activation energy of the process can be inferred from the Arrhenius plot reported in Fig. 2. From the comparison of the blanket film and lines behaviour, two interesting facts can be observed. The first one refers to the shift between the two Arrhenius plots. The second one is that the two sets of experimental data have a different fit slope.

Thermal deterioration processes in blanket and patterned silicide are quite different to each other. The main difference arises from the lateral constraints to the patterned silicide by the mask oxide. In blanket silicide the resistance increase upon annealing comes from the increase of the silicide resistivity [13]. This occurs when surfaces and interfaces limit the mean free path (λ_0) of the electrical carriers. In blanket as reacted silicide λ_0 is much lower than grain size and thickness. Typically, λ_0 at room temperature is 80 nm long [14], while grain size and thickness of our sample are, respectively, 50–200 and 100 nm. A high temperature annealing process changes drastically grain size and shape of the silicide, so that, on average, the silicide thickness close to the grain boundary becomes thinner than λ_0. Taking into account the resistivity dependence on the layer thickness [15], a simple model [13] explains the observed resistance increase versus silicide roughness.

A different situation arises in patterned silicide thin film. The as reacted silicide has a mean grain size of ∼100 nm. The silicide thickness is uniform in the center of the line and is reduced near the oxide edge. High temperature annealing produces a strong interface roughness [16]. The silicide becomes very thin close to the lateral oxide, while it reaches the underlying oxide layer in the center of the strip, as shown in Fig. 3b. At the same time, a large grain growth from ∼100 to ∼200 nm close to the center of the line can be observed. From the plan-view TEM reported in Fig. 3a, an explanation of the annealed strip electrical behavior can be inferred. The main information is that a "lateral" silicide roughness rises after thermal process. Moreover, some unconnected silicide grains are left at the border. The observed change in the strip morphology can be described as a three-dimensional grain growth, preferentially in the middle of the strip, with atomic diffusion from the edges toward the line center. Furthermore, it has been observed that the average transverse area is reduced after annealing because some unconnected silicide islands are left at the border. From the plan view TEM reported in Fig. 3a, a reasonable estimation for the mass loss is ∼20% and a lateral roughness of about 200 nm is found. To check the sensitivity of the resistance to the local observed spread $(A - A_0)$ of the transverse area along the line, we have calculated the strip resistance using the following formula:

$$R = \rho L \int_{A_0 - 2\sigma}^{A_0 + 2\sigma} \frac{1}{A\sqrt{2\pi\sigma}} \exp\left[- \frac{(A - A_0)_2}{2\sigma^2} \right] dA \qquad (1)$$

where σ is the mean square variation of the transverse area and A_0 is the value measured in the as-grown sample (10^{-5} cm^2). Taking into account the linewidth variation ($\sigma = 200$ nm) and the mass loss ($A_0 = (1-0.2) \times 10^{-5}$ cm^2) observed, it was possible to obtain a resistance value very close to the experimental one with our model.

From a microscopic point of view, the final configuration results from different thermally activated processes, having different activation energies. The slowest physical process coming into play will attend the deterioration process. The two activation energies found (Fig.2) are totally different, being higher for the blanket silicide than for the patterned one. In the former E_a is equal to 5.0 ± 0.2 eV, while in the latter it is 3.5 ± 0.4 eV. This means that different processes attend the evolution of the silicide morphology. In blanket silicide layers, grain growth and grooving at the grain boundary occurs, following the isothermal annealing time [13,17]. This agglomeration process starts at grain boundary where break of the Co–Si bond in the CoSi$_2$ molecules is favored. Cobalt flows towards the interface and reacts with the underlying silicon, leading to grain growth. Instead, Si atoms coming from the Co–Si couple at the G.B. regrows on the silicon underlying grains. In this way, silicon grains are able to increase in dimension, contributing to the silicide deterioration at the G.B. [7]. Every step of the deterioration process is characterized by particular activation energy. In cobalt silicide, 5 eV is the energy required to break the Co–Si bond at the grain boundary [13] The other microscopic processes coming into play have much lower activation energy. Then, this result suggests that the CoSi$_2$ film itself rather than the substrate microstructure plays the dominant role in the thermal stability. Therefore, in this class of sample, surface film and substrate are uncoupled. This condition is comparable to that observed on crystalline silicon, where an activation energy of 5 eV is found [17].

In order to support the previous assertion, we have realized a particular experimental condition where the substrate has a notable impact. We replaced the silicon implant, performed prior to Co deposition, with high dose BF$_2$ ion implant. The ion energy was chosen in such a way that the peak of the impurity after the silicide reaction is found close to the silicide/silicon interface. It has been demonstrated [10] that this is the best experimental condition in order to improve the silicide

thermal stability. If we assume that the fluorine buffer model [18] could be used also in our sample, we expect a better thermal stability in the BF$_2$ than in the Si implanted samples. Following the standard procedure, resistance measurements in the temperature range between 950 and 1100°C are performed. Again, a linear increase of the sheet resistance versus annealing time was found. From the Arrhenius plot shown in fig.2, an activation energy of 5.7 ± 0.8 eV has been found. This value is comparable to that of the silicon-implanted sample, within the experimental errors. Therefore, in these experimental conditions the interface fluorine atoms perform no improvement of stability, so the only silicide layer comes into play.

The microscopic description of the process is different on the lines. Upon annealing, the increase of the silicide grain diameter in the center of the line becomes clear looking at the TEM plan view and cross-section reported in Fig. 3a,b, respectively. This means that during the thermal process the CoSi$_2$ molecules break and the Co atoms flow preferentially toward the strip center. In this region these atoms react with the substrate producing the large grains previously shown. During this process the free silicon atoms, produced by the break of the Co–Si bonds close to the grain boundaries, migrate to the silicon/silicide interface and arranged themselves epitaxially on the polysilicon grains of the substrate. The final result, shown in Fig. 3b, is the formation of large silicon grains with a low defect density at the line edge [3]. Then, it is clear that the epitaxial silicon growth is the slowest process and is the one limiting the CoSi$_2$ deterioration in the lines.

The reason for this difference between the activation energy in the blanket and in the laterally limited films is not completely clear and further investigations are needed. The main difference observed in the two types of samples is the substrate morphology. In fact, while the blanket sample has a polysilicon substrate with grains having a large defect density (not shown), the substrate of the lines is almost free of dislocations and twins. Probably, this good crystallinity of the substrate produces a reduction of the epitaxial growth rate and then the process that limits the deterioration of the line becomes the silicon epitaxial growth. We think that the atomic local equilibrium at the edges, after the break of the Co–Si bonds, depends on the boundary conditions. A substrate with many defects contributes considerably to lowering the free Si atoms concentration coming from the break in the silicide. Then, the equilibrium between CoSi$_2$ and Co + Si atoms shifts toward the former and the break/formation of the bonds limits the process. Instead, a substrate having a good crystallinity slows down the Si flow from the silicide, because of the low growth rate of these particular silicon grains. Therefore, this process limits the silicide deterioration and becomes the driving one.

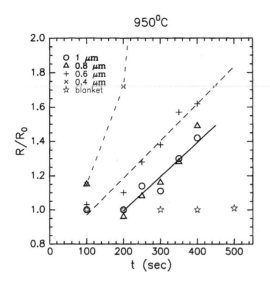

Fig. 4. Normalised resistances increase versus annealing time for lines reacted on (001) silicon. The effect of the lateral dimensional scaling on the silicide thermal stability is evident.

3.2. Lateral dimension scaling on (001) Si

To show that the effect of lateral dimension scaling on the thermal stability is a general effect and does not depend on the particular substrate, the $CoSi_2$ high temperature stability has been studied in lines and in blanket regions on a (001) substrate. Resistance measurements have been performed in the temperature range between 900 and 1050°C. Nominal stripe width goes from 1 to 0.4 µm but it was found that the real ones were ~0.2 µm wider because of Co reaction under the mask oxide. TEM cross-section analysis (not reported) also indicate a silicide thickness of 150 nm and the obtained $CoSi_2$ film resistivity was of about 15 µΩ cm. In Fig. 4 the thermal stability of both patterned and blanket silicide has been compared at 950°C. Resistance measurements have been normalized to that of as reacted samples and, within the time window probed, the blanket silicide is systematically the most stable. In this kind of sample, no resistance increase has been detected. Instead, in the 1 µm wide stripes the resistance starts to increase from 200 s anneal with a degradation rate of $\sim 2 \times 10^{-3}$ s^{-1}. After an initial change of resistance, depending on the silicide morphology arrangement, also the resistance of the strip 0.8 µm wide increases linearly with the same rate. In the 0.6 µm strip, the delay time, spent before

the silicide degradation starts, is reduced to 100 s. However, the rate is similar to the former. A totally different behavior in the narrowest lines has been found. The delay time is further lowered, unlike the deterioration rate which becomes higher. This trend is the same at different temperatures, from 900 to 1050°C. The narrower the strip, the quicker the degradation process [18]. Moreover, resistance measurements on the strip with the same width (1 µm) performed on $CoSi_2/(001)$ silicon (fig.4) and $CoSi_2$/polycrystalline silicon (fig.1) can be compared. The anneals were performed at the same temperature and a degradation rate lower in the former than in the latter has been found.

4. Conclusions

The lateral dimension scaling effect on the $CoSi_2$ thermal stability was demonstrated. The difference between activation energies found for blanket and patterned silicide on polycrystalline silicon was explained. The first one (5 eV) is related to either breaking or formation of the Co–Si bond, the second one (3.5 eV) to the underlying silicon grain growth Moreover, thermal stability dependence on line width on crystalline silicon was proven.

References

[1] Lasky JB, Nakos JS, Cain OJ, Geiss PJ. IEEE Trans Electron Devices 1991;38:262.

[2] Colgan EG, Gambino JP, Hong QZ. Mater Sci Eng 1996;R16:43.

[3] Hong QZ, Hong SQ, D'Heurle FM, Harper JME. Thin Solid Films 1994;253:479.

[4] Wang QF, Osborn CM, Smith PL, Canovai CA, McQuire GE. J Electrochem Soc 1993;140:200.

[5] Karlin TE, Zhang SL, Rydén KH, Nygren S, Östling M, D'Heurle FM. Appl Surf Sci 1993;73:277.

[6] Osburn CM, Tsai JY, Wang QF, Rose J, Cowen A. J Electrochem Soc 1993;140(12):3660.

[7] Chen WM, Banerjee SK, Lee JC. Appl Phys Lett 1994;64(12):1505.

[8] Schreutelkamp RJ, Deweerdt B, Verbeeck R, Maex K. Micoelectron Eng 1992;19:665.

[9] Wang QF, Tsai JY, Osburn CM, Chapman R, McGuire GE. Appl Phys Lett 1992;61(24):2920.

[10] Chen BS, Chen MC. J Appl Phys 1993;74(2):1035.

[11] Wang QF, Osburn CM, Smith PL, Canovai CA, McGuire GE. J Electrochem Soc 1993;140(1):200.

[12] Lasky JB, Nakos JS, Cain OJ, Geiss PJ. IEEE Trans. Electron Devices 1991;38:2.

[13] La Via F, Alberti A, Raineri V, Ravesi S, Rimini E. J Vac Sci Technol 1998;16:3.

[14] Hensel JC, Tung RT, Poate JM, Unterwald F. Phys Rev Lett 1985;54(16):1840.

[15] Ohring M. The material science of thin film. San Diego: Academic. 1992. p. 459.

[16] Norström H, Maex K, Romano-Rodriguez A, Vanhellemont J, Van den Hove L. Microelectron Eng 1991;14:327.

[17] Jiang H, Osburn CM, Xiao ZG, McGuire G, Rozgonyi GA. J Electron Soc 1992;139(1):211.

[18] Tsui BY, Wu TS, Chen MC. IEEE Trans. Electron Devices 1993;40(1):54.

PERGAMON

Solid-State Electronics 43 (1999) 1045–1049

SOLID-STATE
ELECTRONICS

An investigation into the performance of diffusion barrier materials against copper diffusion using metal-oxide-semiconductor (MOS) capacitor structures

Vee S.C. Len*, R.E. Hurley, N. McCusker, D.W. McNeill, B.M. Armstrong, H.S. Gamble

Department of Electrical and Electronic Engineering, The Queen's University of Belfast, Belfast, BT7 1NN, Northern Ireland, UK

Abstract

Cu/SiO$_2$/Si capacitors were fabricated, with and without barrier layers between the copper and the oxide, and the dielectric properties of the 100 nm thermal SiO$_2$ layers monitored after high temperature stressing. Film properties were also examined by X-ray diffraction spectroscopy (XRD) and atomic force microscopy (AFM). Diffusion barriers of Ti, TiN, TiN/Ti, Ta and TaN/Ta were assessed for thermal stability and ability to prevent Cu diffusion. These evaluations indicated that a 10 nm PVD TiN film is a good barrier against Cu diffusion up to 550°C, whilst the addition of a thin Ti layer allows the TiN barrier to withstand a 600°C 60 s anneal. Ta and its nitrides were assessed and found to fail at temperatures as low as 400°C. © 1999 Elsevier Science Ltd. All rights reserved.

1. Introduction

Copper-based interconnects are expected to be in use for ULSI fabrication after the year 2000. Copper is known to diffuse quickly through dielectric layers and silicon. Thus, it is essential to introduce a diffusion barrier between the Cu metallisation and the dielectric. An ideal barrier must be chemically and thermodynamically stable, without increasing the overall resistivity [1,2].

Contradictory reports exist in the literature concerning the diffusion of copper into dielectric. McBrayer et al. [3] have shown diffusion of copper into SiO$_2$ after a 400°C anneal in both N$_2$ and N$_2$/H$_2$ ambient. Other workers found that Cu migration into SiO$_2$ was evi-

dent after a 300°C anneal [4,5]. On the contrary, there was little or no Cu migration into the SiO$_2$ even after annealing at 450–500°C if the thermal stressing was carried out under vacuum [3,4] or by encapsulating the structure to prevent Cu oxidation [6].

Titanium nitride (TiN) has been used to prevent interdiffusion of Al and its alloys, and a sub-100 nm TiN$_x$ barrier has been found to prevent Cu diffusion between 450 and 650°C [7–9]. Earlier investigations into tantalum (Ta) and its nitride (TaN) reported that Ta was able to prevent Cu diffusion up to 550–750°C, and TaN films were effective up to 650–700°C [10–12].

The contradictory results above may be due to variations in the reactive sputtering conditions used for the Ti and Ta nitride. In this paper, we report on the effects of high temperature stresses on Cu MOS structures, and evaluate the effectiveness of barriers such as thin Ti, Ta and their nitrides and then compare the failure temperature and mechanism for each case.

---

* Corresponding author. Tel.: + 44-1232-274340; fax: + 44-1232-667023.

E-mail address: vlsc@hotmail.com (V.S.C. Len)

0038-1101/99/$ - see front matter © 1999 Elsevier Science Ltd. All rights reserved.
PII: S0038-1101(99)00022-2

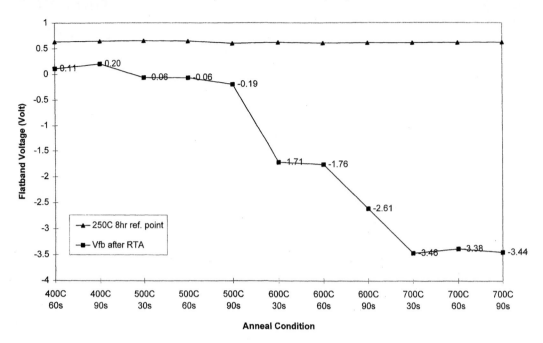

Fig. 1. Flatband voltage (V_{FB}) of Cu/SiO$_2$/Si MOS capacitors vs annealing temperature of 30–90 s RTA in N$_2$ ambient.

2. Experimental details

2.1. Cu/SiO₂/Si MOSC

Cu gate MOS capacitors were fabricated on 100 mm diameter < 100 > n-Si wafers. After SC1 cleaning, a 110 nm thermal SiO$_2$ layer was grown. Cu films, 150 nm, were sputter deposited using a Cu target (99.998% pure) in a dc magnetron sputtering system at a deposition rate of 150 nm/min at 400 W. The copper films were patterned to form the gate electrode. Aluminium was used to make ohmic contact to the back of the wafer. On one set of Cu/oxide structures, the gates were covered with 150 nm thick HSG-2209 S-R7 Spin On Glass (SOG), to prevent oxidation during annealing.

2.2. Cu/diffusion barrier/SiO₂/Si MOSC

The oxide was covered with barrier layers consisting of either 10 nm TiN or 10 nm TiN on 5 nm Ti deposited by reactive sputtering in the dc magnetron sputtering system at a rate of 10 nm/min. A 150 nm Cu layer was then deposited without breaking the vacuum, and the metal stack patterned into 1.2 mm diameter dots

by lift-off. Alternatively, barrier layers of 50 nm Ta and TaN/Ta films were reactively sputtered using a Ta target (99.99% pure) in an Edwards RF magnetron sputtering system at a rate of 40 nm/min at 400 W in Ar and Ar/N$_2$ ambient. Samples were transferred immediately into the dc magnetron sputtering system for deposition of 150 nm Cu. The composition of both TiN and TaN films was controlled by an optical emission spectroscope (OES).

Thermal stressing was performed in a Sitesa rapid thermal anneal (RTA) system from 400 to 800°C for 30–90 s in N$_2$ ambient, after SOG passivation. After the anneals, contact windows were opened in the SOG to allow subsequent electrical measurements. All samples were given a 8 h 250°C anneal in N$_2$ before further thermal stressing, as this proved optimum to remove radiation damage.

Electrical assessment of the copper penetration was done via C–V analysis before and after each annealing schedule. Reactions and interdiffusions between the Cu/barrier/oxide/Si structures were analysed using X-ray diffraction spectroscopy (XRD). Microstructure and film morphology were observed using a Nanoscope III atomic force microscopy (AFM) in tapping mode.

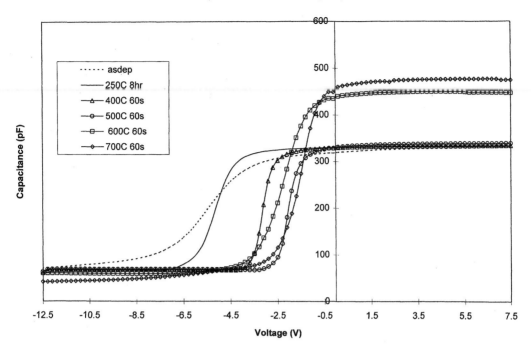

Fig. 2. C–V characteristics of Cu/Ti/SiO$_2$/Si MOS structures after 400–700°C 60 s RTA.

3. Experimental results and discussion

3.1. Cu on oxide

Fig. 1 shows the shifts in flatband voltage (V_{FB}) for different anneal times and temperatures for Cu/SiO$_2$/Si MOS capacitors. Shifts in V_{FB} indicate that copper diffusion increases with annealing temperature from 0.6 V (after 250°C 8 h anneal) to −3.4 V (after 700°C RTA). This large shift in V_{FB} can be attributed to the movement of positive Cu ions into the dielectric. It is noted that Cu migration through the SiO$_2$ is a sensitive function of annealing temperature.

In the SOG-passivated Cu/SiO$_2$/Si system, Cu diffusion into the dielectric was evident at a slower rate. Effects of passivation appear marginal at lower anneal temperatures, but in the high temperature range from 600 to 800°C, changes in V_{FB} values of 1 V were observed. It is postulated that Cu-oxide on the gate surface releases Cu ions readily which diffuse into the SiO$_2$ at a faster rate.

3.2. The barrier performance of Ti and TiN films

Initial checks were carried out on control barrier/oxide/Si MOS capacitors to determine any instabilities so that changes of the Cu gate/barrier/oxide/Si MOS structures could be properly evaluated.

3.2.1. Cu/Ti/SiO$_2$/Si

C–V characteristics of Cu/Ti/SiO$_2$/Si structure as depicted in Fig. 2 reveal that at RTA temperatures from 600 to 700°C, the maximum capacitance has increased compared to the as-deposited and 400 to 500°C annealed capacitors. This implies that the Ti barrier layer has chemically reacted with SiO$_2$ to form Ti–O [13]. The high dielectric constant of TiO$_x$ results in the higher value of capacitance. Ti may be used as a sacrificial layer to prevent Cu diffusion for a limited length of temperature and time.

The formation of different phases of Cu and Ti compounds has been reported to occur at temperature as low as 350 and up to 475°C. A sandwich layer is necessary in order to prevent the intermixing of the Cu–Ti and Ti–SiO$_2$ during the high temperature treatment.

3.2.2. Cu/TiN/SiO$_2$/Si

For a Cu/10 nm TiN/SiO$_2$/Si MOS structure, no obvious copper ions were detected after a 550°C 60 s RTA. Upon higher annealing temperatures, V_{FB} shifted from 0.33 V (550°C) to −0.43 V (800°C). AFM photographs of reference TiN samples are illustrated

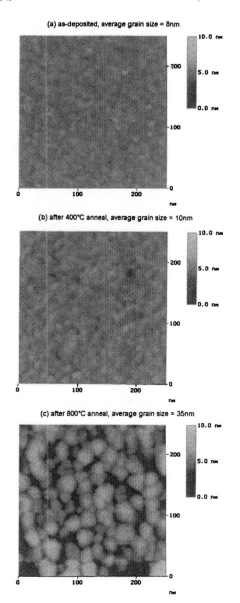

(a) as-deposited, average grain size = 8nm

(b) after 400°C anneal, average grain size = 10nm

(c) after 800°C anneal, average grain size = 35nm

Fig. 3. AFM scans of 10 nm TiN film (a) as-deposited, average grain size = 8 nm; (b) after 400°C anneal, average grain size = 10 nm and (c) after 800°C anneal, average grain size = 35 nm.

in Fig. 3. Grain sizes vary from 8 nm as-deposited to 35 nm after a 800°C anneal. The fine grain structures could provide diffusion paths for Cu to penetrate into the oxide via the grain boundaries. Glancing X-ray diffraction patterns as shown in Fig. 4(a)–(d) document

the reaction mechanism of the Cu/TiN barrier after each anneal schedule. Fig. 4(a),(b) shows no reaction between the as-deposited Cu and TiN films at 400°C. No peaks corresponding to TiN were found but this might be attributed to the sensitivity of the XRD to a 10 nm thick layer. Fig. 4(c),(d) shows a reaction occurring with a peak at around 36.5° after an 800°C anneal which may be attributed to Cu_3Si or CuO.

3.2.3. $Cu/TiN/Ti/SiO_2/Si$

For Cu/15 nm TiN/5 nm Ti/SiO_2/Si MOS structures, the Ti serves as a sacrificial adhesion layer. The TiN serves as an additional diffusion barrier against Cu diffusion. This bilayer system was effective for RTA up to 600°C. Upon higher annealing temperatures, V_{FB} shifted from -0.1 V (600°C) to -0.5 V (700°C).

3.3. The barrier performance of Ta and TaN films

3.3.1. $Cu/50$ nm $Ta/SiO_2/Si$

In this work, early failure of 50 nm Ta barriers was observed after a 60 s 400°C anneal. The failure mechanism in this situation is mainly due to Cu movement via the polycrystalline grain boundaries of the Ta film.

Clevenger et al. [14] demonstrated high vacuum deposited Ta barrier failure at 560–630°C, and ultra-high vacuum Ta barrier failure from 310 to 630°C. In the latter case, the Ta barrier is too pure and thus, to improve performance, its grain boundaries were stuffed with N_2 during Ta deposition to form a TaN barrier layer.

3.3.2. $Cu/50$ nm and 80 nm $TaN/5$ nm $Ta/SiO_2/Si$

Cu/50 and 80 nm TaN/5 nm Ta/SiO_2/Si MOS capacitor structures were fabricated using the same deposition time and conditions, except that the Ar:N_2 gas ratios were 1:0.7 and 1:0.4, respectively. Initial tests were not satisfactory as the TaN films were highly stressed and delamination occurred. A thin Ta layer was introduced to enhance the adhesion. C–V measurements indicated that both 50 and 80 nm thick TaN layers failed to prevent Cu penetration into the dielectric at 400°C. It was found that the thicker TaN barrier was more resistive to Cu diffusion than the thinner layer, exhibiting a 1 V reduction in V_{FB} shift from 500 to 700°C.

4. Conclusion

Thermal stabilities of TiN and TaN layers as diffusion barrier between Cu and SiO_2 on Si MOS structures were studied by C–V measurements. Without the incorporation of a barrier layer, Cu diffuses at a relatively fast rate into the dielectric under the influence of thermal anneals. The effects of Cu passivation against

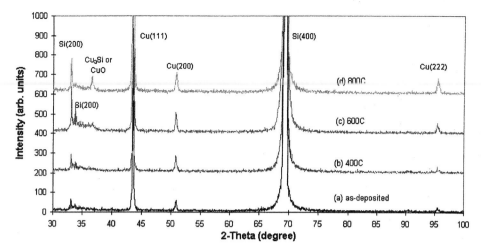

Fig. 4. (a)–(d) shows the glancing X-ray diffraction patterns for the Cu/TiN/SiO$_2$/Si MOS structures in various stages of annealing for 60 s: (a) as-deposited; (b) 400°C; (c) 600°C and (d) 800°C 60 s RTA.

oxidation during subsequent anneals at higher temperatures is more prominent at temperatures above 600°C. A 10 nm thick TiN barrier was found to be effective against Cu diffusion up to 550°C RTA. The addition of 5 nm Ti used as an adhesion cum sacrificial layer to a 15 nm TiN barrier was able to impede Cu penetration during a 600°C RTA. The quality and performance of Ta and its nitrides as Cu barrier were not satisfactory.

Acknowledgements

The authors would like to acknowledge S. H. Raza for his assistance in device fabrication, Geology X-ray laboratory for the XRD measurements and R. Turner for the AFM scans. The authors are grateful to Seagate Technology (NI) Ltd for their support.

References

[1] Nicolet MA. Ternary amorphous metallic thin-films as diffusion-barriers for Cu metallization. Appl Surf Sci 1995;91:269.

[2] Wang SQ. Copper diffusion into silicon and drift through silicon dioxide. MRS Bulletin 1994;19:30.

[3] McBrayer JD, Swanson RM, Sigmon TW. Diffusion of metals in silicon dioxide. J Electrochem Soc 1986;133:1242.

[4] Shacham-Diamand Y, Dedhia A, Hoffstetter D, Oldham WG. Copper transport in thermal SiO$_2$. J Electrochem Soc 1993;140:2427.

[5] Chiou JC, Wang HI, Chen MC. Dielectric degradation of Cu/SiO$_2$/Si structure during thermal annealing. J Electrochem Soc 1996;143:990.

[6] Pai PL, Ting CH. Copper as the future interconnection material. In: Proceedings of VMIC Conference, 1989. p. 258.

[7] Olowolafe JO, Li J, Mayer JW. Interactions of Cu with CoSi$_2$, CrSi$_2$ and TiSi$_2$ with and without TiN$_x$ barrier layers. J Appl Phys 1990;68:6207.

[8] Park KC, Kim KB. Effect of annealing of titanium nitride on the diffusion barrier property in Cu metallization. J Electrochem Soc 1995;142:3109.

[9] Torres J, Mermet JL, Madar R, Grean C, Gessner T, Bertz A, Hasse W, Plotner M, Binder F, Save D. Copper-based metallization for ULSI circuits. Microelectronic Eng 1996;34:119.

[10] Holloway K, Fryer PM, Cabral C, Harper JME, Bailey PJ, Kelleher KH. Tantalum as a diffusion barrier between copper and silicon – failure mechanism and effect of nitrogen additions. J Appl Phys 1992;71:5433.

[11] Stavrev M, Wenzel C, Moller A, Drescher K. Sputtering of tantalum-based diffusion-barriers in Si/Cu metallization – effects of gas-pressure and composition. Appl Surf Sci 1995;91:257.

[12] Oku T, Kawakami E, Uekubo M, Takahiro K, Yamaguchi S, Murakami M. Diffusion barrier property of TaN between Si and Cu. Appl Surf Sci 1996,99.265.

[13] Russell SW. Reaction-kinetics in the Ti/SiO$_2$ system and Ti thickness dependence on reaction-rate. J Appl Phys 1994;76:257.

[14] Clevenger LA, Bojarczuk NA, Holloway K, Harper JME, Cabral Jr C, Schad RG, Cardone F, Stolt L. Comparison of high-vacuum and ultrahigh-vacuum tantalum diffusion barrier performance against copper penetration. J Appl Phys 1993;73:300.

PERGAMON

Solid-State Electronics 43 (1999) 1051–1054

**SOLID-STATE
ELECTRONICS**

Silicide reaction of Co with $Si_{0.999}C_{0.001}$

S. Teichert [a,*], M. Falke [a], H. Giesler [a], G. Beddies [a], H.-J. Hinneberg [a],
G. Lippert [b], J. Griesche [b], H.J. Osten [b]

[a] *Institute of Physics, Chemnitz University of Technology, D-09107 Chemnitz, Germany*
[b] *Institute for Semiconductor Physics, Walter-Korsing-Str. 2, D-15230 Frankfurt (Oder), Germany*

Received 11 August 1998; received in revised form 17 December 1998; accepted 21 January 1999

Abstract

The disilicide of Co is a promising metallic material for the submicron technology with the advantage of a low specific resistivity as well as of a low preparation temperature. A further downscaling in the Si technology requires in addition to lower process temperatures an effective conservation of dopant distributions. One method for trapping doping profiles is the admixture of a small concentration of carbon to the Si. This paper reports on the Co silicide formation on $Si_{0.999}C_{0.001}$(001) substrate layers. The Co films were deposited by e-beam evaporation under UHV conditions. To investigate the reaction of the Co films with the substrate layers performed by rapid thermal annealing the samples were characterized by Rutherford backscattering and measuring the electrical resistance at room temperature. Compared with Si(001) a small increase of the thermal budget for the $CoSi_2$ formation on $Si_{0.999}C_{0.001}$(001) has been found. For samples annealed at 650°C for 30 s the resistivity in dependence on temperature was determined. The analysis of this data using the Bloch–Grüneisen equation reveals a good electrical quality of the prepared $CoSi_2$ films. © 1999 Elsevier Science Ltd. All rights reserved.

1. Introduction

Silicides are widely-used materials for different process types in the production of electronic devices. Whereas $TiSi_2$ at present is commonly used in applications $CoSi_2$ has clear advantages in the future downscaling of microelectronic devices [1]. The submicron Si technology requires among other things the lowering of process temperatures as well as the effective conservation of the dopant distributions. A possible solution of the latter demand is the admixture of a small amount of carbon to the Si substrate. Recently, it has been shown [2] that this carbon can effectively reduce

the outdiffusion of dopants during thermal processing. For the mentioned reasons the reaction of Co with a C containing substrate layer is of great interest. In this work we report on the investigation of the reaction of thin Co films with $Si_{0.999}C_{0.001}$(00l) substrate layers.

2. Experimental

For the silicide growth process two different types of substrate layers were grown on p-Si(001) wafers by molecular beam epitaxy (MBE) at a substrate temperature of 500°C, a carbon containing $Si_{0.999}C_{0.001}$ layer and for comparison a pure Si layer. First, a Si buffer, 50 nm in thickness, was deposited followed by the epitaxy of $Si_{0.999}C_{0.001}$ or Si with a thickness of 100 nm in each case. Some physical properties of the grown $Si_{0.999}C_{0.001}$(001) layers have been given in a previous

* Corresponding author. Tel.: +49-371-531-4114; fax: +49-371-531-3077.
E-mail address: teichert@physik.tu-chemnitz.de (S. Teichert)

0038-1101/99/$ - see front matter © 1999 Elsevier Science Ltd. All rights reserved.
PII: S0038-1101(99)00023-4

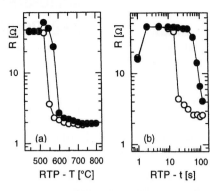

Fig. 1. Four-point resistance measured at room temperature after isochronal (a) and isothermal (b) RTP treatment; open symbols is the Si substrate layer, full symbols is the $Si_{0.999}C_{0.001}$ substrate layer.

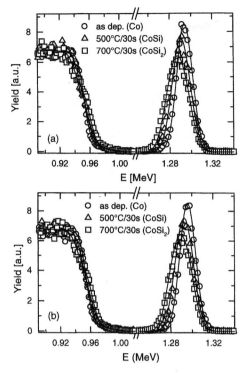

Fig. 2. RBS spectra for films on Si (a) and on $Si_{0.999}C_{0.001}$ (b) substrate layers; Symbols represent measured data and lines RUMP simulations.

paper [3]. A comprehensive description of the UHV equipment as well as of the deposition conditions used in the experiment can be found in [4]. In a second UHV equipment the Co was deposited by e-beam evaporation onto the substrate layers. Prior to this deposition step the wafers were cleaned with a buffered HF solution. During the deposition the substrate temperature was kept below 100°C and the pressure below 10^{-9} Torr.

In all experiments a Co film, 7.2 nm in thickness, was deposited in order to receive 25 nm of $CoSi_2$ after the complete silicide reaction with the substrate layer. The subsequent annealing of the samples was performed in a rapid thermal processor (RTP) in flowing high purity N_2. Series of samples annealed under isochronal ($t = 30$ s) as well as under isothermal ($T = 550°C$) conditions were prepared and analyzed in dependence on the RTP treatment by four-point resistance measurements at room temperature.

Rutherford backscattering spectroscopy (RBS) has been applied to obtain information on the changes in the composition during silicide formation The RBS spectra were taken using an incident beam of 1.7 MeV He^+ ions and a detector angle of 170° and were evaluated by RUMP simulations. The electrical resistivity of the disilicide films was studied by measurement of the resistance in the temperature range from 4.2 to 300 K using a conventional He cryostat. For this measurements the samples were structured by a reactive ion etching process [5].

3. Results and discussion

The evolution of the silicide growth starting with a Co film on Si passing through intermediate stages

when Co_2Si or CoSi occur and ending with the $CoSi_2$ phase can be monitored under certain assumptions by the measurement of the room temperature resistance after different thermal treatments. These assumptions include, that always closed films exist which can be described by the known resistivities of the materials involved and by their thickness given by the total amount of Co kept constant in our experiments. Therefore it is useful directly to compare the room-temperature resistance of different samples. In case of the Co/Si system a rough estimate considering thickness changes [6] as well as the resistivity changes [7] shows that Co_2Si and CoSi can not be distinguished, a given amount of Co will cause about the same resistance for either of the phases. In contrast to that, the transition to $CoSi_2$ is observable by these measurements, a strong decrease of the resistance is expected. Fig. 1 shows the discussed evolution of the resistance of sample series measured at room temperature. The expected resistance drop is clearly visible in case of isochronal (Fig. 1a) as well as isothermal (Fig. 1b) RTP treatment suggesting the transformation to $CoSi_2$ for both sample types. The most important result is the

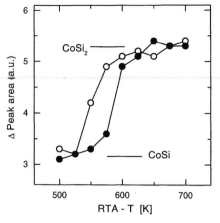
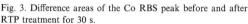

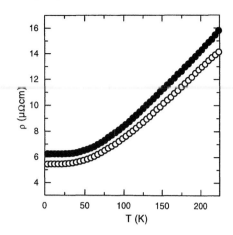

Fig. 3. Difference areas of the Co RBS peak before and after RTP treatment for 30 s.

Fig. 4. Resistivity vs. temperature; open symbols represent Si substrate layer and full symbols $Si_{0.999}C_{0.001}$ substrate layer.

shift of the resistance drop by about 50 K to higher RTP temperatures at 30 s annealing time for the C containing substrate layer compared to pure Si. In case of isothermal heat treatment at 550°C an annealing time of 20 and 120 s is required for Si and $Si_{0.999}C_{0.001}$, respectively, to remarkably decrease the resistance. Both results mean that there exists a certain increase of the thermal budget necessary for the silicide reaction on $Si_{0.999}C_{0.001}$ with respect to the reaction on Si. Moreover Fig. 1b demonstrates indicated by the relatively high resistance difference found after RTP treatment for 120 s that not all of the Co has reacted to $CoSi_2$ in case of the $Si_{0.999}C_{0.001}$ substrate layer for this annealing time. Fig. 1a shows that for isochronal RTP treatment at temperatures higher than 600°C the resistance of both substrate layers approaches. However, the resistance of samples grown on $Si_{0.999}C_{0.001}$ substrate layers has always been found to be a little higher compared to samples on Si annealed under the same conditions. Even at annealing temperatures of 800°C there remains a small resistance difference.

RBS measurements were performed in order to prove the above mentioned phase formation sequence for samples annealed isochronally in the temperature range between 500 and 700°C. In Fig. 2a and b, the RBS spectra are plotted for films on $Si_{0.999}C_{0.001}$ and on Si, in each case for three samples: after deposition of Co, after RTP treatment at 500°C and after 700°C. The comparison of the spectra taken from samples on different substrate layers after the same annealing treatment demonstrates, that for the selected reaction stages there is nearly no influence of the incorporated carbon. However, this means only, that for these particular temperatures the process of the silicide reaction

is progressed nearly to the same degree for both substrate types. RUMP simulations of the RBS spectra were performed and suggest for both sample types the existence of mainly CoSi at 500°C and of $CoSi_2$ for temperatures higher than 700°C. All simulations confirm that for all measured samples the total Co amount given by the deposition thickness remains constant. The transformation from CoSi to $CoSi_2$ is accompanied by a spatial broadening of the Co distribution and by a decrease of the mean Co concentration in the silicide film, causing the broadening and decreasing of the Co signal in the RBS spectrum. To get a quantitative measure for the progress of the silicide reaction deduced from the RBS spectra the following procedure has been applied. First, a difference spectrum was calculated using the spectra of the same sample taken before and after thermal treatment. Then the absolute value of the area of the Co structure in this spectrum was calculated and taken as the measure in question. In Fig. 3 there is shown the plot of these difference areas in dependence on the temperature of the isochronal RTP treatment. For comparison the expected values for pure CoSi and $CoSi_2$ films located at the sample surface and simulated by RUMP have been inserted into the plot. Again, the most important result is the clearly visible shift of the disilicide formation temperature to higher values due to the C incorporated in the substrate layer. This result of the RBS measurements is in remarkably good agreement with the corresponding results of four-point resistance measurements at room temperature (Fig. 1a).

In order to analyze the electrical transport properties in more detail temperature dependent measurements of the resistance were performed for samples annealed at

Table 1
Fitting parameters for $CoSi_2$ films obtained by using Eq. (1) with n = 3. The resistivity $\rho(T = 300\ K)$ was calculated using these results

Substrate layer	ρ_0 [$\mu\Omega cm$]	ρ_1 [$\mu\Omega cm$]	Θ_D [K]	$\rho(T = 300\ K)$ [$\mu\Omega cm$]
Si	5.4	51	525	18.4
$Si_{0.999}C_{0.001}$	6.2	58	545	20.3

650°C for 30 s. In Fig. 4 is shown the resistivity, calculated by using the intended thickness of 25 nm in dependence on the temperature for both types of substrate layers. The resistivities of the samples show the typical metallic behavior with a nearly linear increase of the resistivity due to electron–phonon scattering at higher temperatures and a constant residual resistivity at low temperatures due to scattering at static imperfections in the material. At temperatures higher than 230 K (not shown) the resistivity deviates from the linear behavior due to increased shunting by the substrate. The Bloch–Grüneisen equation [8]

$$\rho(T) = \rho_0 + \rho_{Ph}$$

$$= \rho_0 + \rho_1 \cdot \left(\frac{T}{\Theta_D}\right)^n \cdot \int_0^{\Theta_D/T} \frac{z^n \cdot e^z}{(e^z - 1)^2} dz \qquad (1)$$

has been used to describe the measurements with the exponents $n = 3$ or $n = 5$ and with free parameters Θ_D, ρ_0 and ρ_1. The best fitting has been obtained in case of $n = 3$ pointing to phonon mediated interband s–d-scattering of electrons [9,10]. This result coincides with previously reported results on $CoSi_2$ samples [11]. The fitting parameters have been summarized in Table 1. The Debye temperature Θ_D agrees well with the known value of 510 K for $CoSi_2$ [12]. However, the residual resistivity for the two samples is relatively high with respect to the typically reported values of about 2 $\mu\Omega$ cm for $CoSi_2$ films. This difference can be explained taking into account the low annealing temperature and the short annealing time which might be responsible for a variety of imperfections such as point defects or grain boundaries causing an increased residual resistivity.

4. Summary

The reaction of Co with $Si_{0.999}C_{0.001}$ has been investigated. A small increase of the thermal budget necessary for disilicide formation was observed for the C containing substrate layer compared to a pure Si substrate layer. RBS measurements show that after an RTP treatment at 700°C for 30 s the Co is transformed into the $CoSi_2$ phase for both types of substrate layers. Measurements of the resistivity in dependence on tem-

perature reveal the metallic behavior of the grown silicide layers. $CoSi_2$ films formed on $Si_{0.999}C_{0.001}$ during RTP annealing at 650°C for 30 s have a higher resistance than corresponding films prepared on pure Si. This difference may be explained by morphological as well as structural differences between these films. However, the experiments have shown altogether, that $CoSi_2$ films of good quality could be obtained within reasonable process parameters using substrate layers of the composition $Si_{0.999}C_{0.001}$. Future projects will concern the investigation of the contact resistance and the behavior of dopant distributions during $CoSi_2$ formation.

Acknowledgements

We thank R. Groetzschel from the FZ Rossendorf for the support at RBS measurements.

References

[1] Maex K. Mater Sci Eng Rep 1993;R11:53.
[2] Osten HJ, Lippert G, Knoll D, Barth R, Heinemann B, Ridcker H, Schley P. IEDM Techn Digest, 1997:803.
[3] Osten HJ, Kim M, Pressel K, Zaumseil P. J Appl Phys 1996;80:6711.
[4] Kim M, Lippert G, Osten HJ. J Appl Phys 1996;80:5748.
[5] Beddies G, Falke M, Teichert S, Gebhardt B, Hinneberg H-J. E-MRS 1998 Spring Meeting, Symp. N. Appl Surf Sci, in press.
[6] M Ostling, Zaring C. In: Maex K, van Rossum M, editors. Properties of metal silicides. London, UK: INSPEC, 1995. p. 15.
[7] Gottlieb U, Nava F, Affronte M, Laborde O, Madar R. In: Maex K, van Rossum M, editors. Properties of metal silicides. London, UK: INSPEC, 1995. p. 189.
[8] Ziman JM. Electrons and phonons. Oxford University Press, 1967.
[9] Webb GW. Phys Rev 1969;181:1127.
[10] Sivaram S, Ficarola PJ, Cadien KC. J Appl Phys 1985;58:1314.
[11] Nava F, Tu KN, Thomas O, Senateur JP, Madar R, Borghesi A, Guizzettl G, Gottheb U, Laborde O, Bisi O. Mat Sci Rep 1993;9:141.
[12] Briggs A, Thomas O, Madar R, Senateur JP. Appl Surf Sci 1991;53:240.

PERGAMON

Solid-State Electronics 43 (1999) 1055–1061

SOLID-STATE ELECTRONICS

Characteristics of sputter-deposited TiN, ZrB$_2$ and W$_2$B diffusion barriers for advanced metallizations to GaAs

M. Guziewicz[a,*], A. Piotrowska[a], E. Kamińska[a], K. Gołaszewska[a], A. Turos[b], E. Mizera[c], A. Winiarski[d], J. Szade[d]

[a]*Institute of Electron Technology, Al. Lotników 32/46, Warsaw, Poland*
[b]*Institute of Nuclear Studies, Hoza 69, 00-668 Warsaw, Poland*
[c]*Institute of Physics of Polish Academy of Sciences, Al. Lotników 32, Warsaw, Poland*
[d]*University of Silesia, Uniwersytecka 4, 40-007 Katowice, Poland*

Received 9 August 1998; received in revised form 17 November 1998; accepted 21 January 1999

Abstract

The sputter deposition of TiN, ZrB$_2$ and W$_2$B thin films were studied in order to develop the process parameters for advanced metallizations to GaAs devices. Thin films of TiN, ZrB$_2$ and W$_2$B were deposited on both bare and AuZn-metallized (100) GaAs substrate in magnetron sputtering systems. The resistivity and stress in as-deposited thin films were examined and related to deposition conditions. The film microstructure and composition of grown compounds were determined by X-ray diffraction, transmission electron microscopy (TEM) and X-ray photoelectron spectroscopy (XPS). It was observed that metal boride films were polycrystalline or amorphous depending on the sputtering conditions. Diffusion barrier properties were analyzed by Rutherford backscattering spectroscopy (RBS) technics. © 1999 Elsevier Science Ltd. All rights reserved.

1. Introduction

Since semiconductors technology has been extended into the submicron regime the demands a high quality ohmic contact dramatically increased. In the case of contact to GaAs the gold-based metallization is still attractive because Au is corrosion resistant and a low specific contact resistance of 10^{-6} Ω cm^2 can be achieved [1]. During thermal formation of ohmic contact or other thermal processing the gallium arsenide reacts with the gold liberating some amount of arsenic. Because of reduced lateral and vertical dimensions of contact and requirements of high electrical conductance, both the decomposition of semiconductor substrate and the depth of penetration of metallization should be reduced to a minimum. A way for solution of that problem is a formation of diffusion barrier film (BF) [2] interposed between contact metal and top layer. It is expected that BF acts as capping layer to suppress As loss and to inhibit diffusion of gold [3]. From many refractory metal borides and nitrides compounds we chosen TiN, ZrB$_2$ and W$_2$B as material for barrier film.

In this work we report a study of sputter deposited TiN, ZrB$_2$ and W$_2$B BF for Au-based metallization to GaAs. The first purpose was the preparation of thin BF with low resistivity and low stress; the second was the evaluation of barrier properties in p-GaAs/AuZn/BF/Au contact structure. A special attention was paid on the effectiveness of BF for As loss suppression and effectiveness of the compound as diffusion barrier to gold migration.

* Corresponding author.

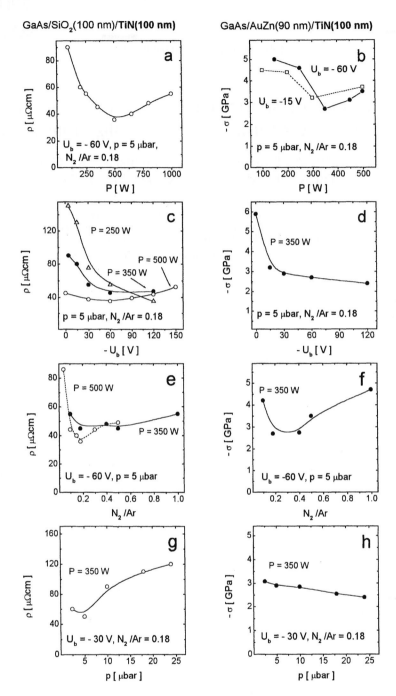

Fig. 1. The dependencies of TiN film resistivity, ρ, and stress, σ, on the sputtering parameters: rf power (a and b), negative bias voltage U_b (c and d), gas flow ratio N_2/Ar (e and f) and total pressure (g and h), respectively. The results concern the film at the thickness of 100 nm.

Table 1
The developed sputtering parameters for TiN, ZrB_2 and W_2B growth and characteristics of the films

Parameter	TiN growth	ZrB_2 growth	W_2B growth	
Power supply	rf	dc	rf	dc
Power discharge: P (W)	350	200	200	200
Substrate bias: U_b (V)	− 60	0	− 60	0
Argon flow: Ar (sccm)	100	30	100	30
Nitrogen flow: N_2 (sccm)	18	–	–	–
Total pressure: p (μbar)	5	5	5	5
Film stress (GPa) (at thickness (nm))	− 1.3 (50)	− 0.75 (60)	− 1.1 (100)	− 0.2 (100)
Result of composition measurement	from XPS	from XPS	from RBS	from RBS
Elements on the bare film (at%)	Ti 23, N 23, O 18, C 36	Zr 15, B 12, O 38, C 35,	–	–
Elements in the film (at%)	Ti 50, N 50	Zr 58, B 42	W 65, B 35	W 55, B 45

2. Experimental conditions

The films of TiN, ZrB_2 and W_2B were deposited by magnetron diode sputtering method in commercial coating systems. The TiN films were prepared by reactive rf sputtering from a Ti target (99.99% purity), in a mixture of argon and nitrogen. W_2B and ZrB_2 layers were deposited from composite target (99.5% purity) powered by rf and/or dc respectively, in Ar atmosphere. Magnetron with 7.5 cm diameter target was water-cooled, target to substrate distance was 6.5 cm.

Both AuZn-metallized and bare (001)GaAs wafers were used for the substrate. The 360 μm thick wafers were cut to the dimension of 10 × 10 mm. Semiinsulating GaAs substrate was used for sheet resistance measurement with four point probe method. Au–Zn metallization of p-GaAs were made by sequential Au(20 nm)/Zn(10 nm)/Au(60 nm) layers evaporation from resistance heated crucibles.

The film thickness was measured by Alphastep 200 system. Stress measurements were performed with Tencor FLX 2320 system, which measures the curvature of sample and calculates the stress according Stoney formula. The microstructure and phase composition of TiN, W_2B and ZrB_2 layers prepared at few set of process parameters were determined both by X-ray diffraction and TEM. The percentage atomic concentration for the film composition was determined by XPS measurements. Composition depth profile of GaAs/AuZn/TiN structure was made via argon ion beam etching after each cycle: milling at 0.5 keV for 5 min, a measurement.

Heat treatments were carried out in conventional furnace under the flow of H_2, for time of 3 min at temperature of 320 or 420°C and by means of rapid thermal annealing in argon atmosphere for ohmic contact formation. Diffusion barrier properties of films were determined by 2 MeV He$^+$ RBS technique and chromium collector method [4].

For mechanical and electrical characterization of barrier films, the dependence of resistivity, ρ, and stress, σ, vs. growth conditions were taken for a film thickness of 100 nm.

3. Results and discussion

3.1. TiN thin films

The silver–gold TiN film grown at low nitrogen flow N_2 of 5 sccm and argon flow Ar of 100 sccm ($N_2/Ar = 0.05$) points out a titanium excess in the film, so higher nitrogen flow has been used. The X-ray diffraction measurements reveal that the TiN layers grown at power of 500 W, bias of −60 V and N_2/Ar from 0.10 to 0.50 on GaAs/Au as a substrate have the polycrystalline structure with single TiN phase and randomly oriented grains. The lattice parameter evaluated from 002 peak is equal 0.424 nm.

The influence of the process parameter like power P, substrate bias U_b, total pressure p, nitrogen flow N_2 on resistivity ρ and stress σ is presented in the Fig. 1. Fig. 1a and b present influence of power on the resistivity and stress σ of TiN film deposited at argon pressure of 5 μbar, N_2/Ar gas ratio of 0.18 and substrate bias of −60 V. The resistivity vs. power reveals a minimum of 36 μΩ cm at power of 500 W. Higher resistivity of TiN film at power of 100 W comes from an impurity species like residual gases, which can be effectively trapped in film at low deposition rate. A minimum of stress (−2.7 GPa) is observed at 350 W. It can be seen in Fig. 1c, that the resistivity vs. bias voltage is enhanced at low bias especially for film grown at power of 250 W. If the bias voltage is increased up to −60 V, TiN resistivity decreases below 60 μΩ cm. It can be connected with reduction of oxygen content in the metallic films due to repelling of oxygen ions from negative biased substrate. For film grown at power of 500 W the resistivity has a flat minimum at −60 V. With increase of bias voltage from 0 V up to −120 V

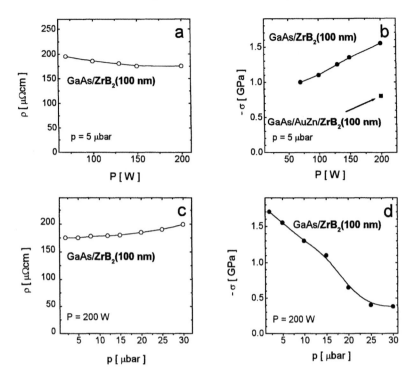

Fig. 2. The resistivity, ρ, and stress, σ, of 100 nm ZrB$_2$ film as a functions of dc power (a and b) and of argon pressure (c and d).

the compressive stress decreases from 5.9 GPa down to 2.4 GPa.

Both resistivity of TiN film vs. pressure, as well as vs. N$_2$/Ar ratio reveals a minima at $p = 5$ μbar and at N$_2$/Ar = 0.18, (Fig. 1e and g), respectively. With increasing pressure from 2 μbar up to 25 μbar the negative stress decreases from 3.5 GPa down to 2.4 GPa (Fig. 1f). Taking into consideration the minimum for resistivity and stress the conditions: $P = 350$ W, $U_b = -60$ V $p = 5$ μbar and N$_2$/Ar = 0.18 were chosen for study barrier properties in contact structure GaAs/AuZn/TiN/Au.

From planar TEM measurements of AuZn(90 nm)/TiN(50 nm) contact structure it follows that the width of TiN grains (below 50 nm) is considerable smaller than width of AuZn grains.

XPS measurements of atomic concentration of elements in the TiN film are listed in Table 1. Ti 2p3/2 and N 1s peaks were registered at binding energy of 454.85 and 397.2 eV, respectively. Such binding energies are very closed to the values obtained by Chourasia and Chopra [5] for TiN compound. After Ar$^+$ milling for 20 min (removing about 10 nm of the top layer) the O 1s peak (at 530.7 eV) and C 1s peak (at 285.0 eV) drooped into background level. N/Ti

ratio of 1.00 ± 0.02 was calculated from the areas under the Ti 2p3/2 and the N 1s peaks. TiN stoichiometry was stable during etching of GaAs/AuZn(90 nm)/TiN(70 nm) contact structure and no spectral changes were observed in Ti 2p3/2, N 1s, Au 4f7/2, Zn 2p3/2 regions indicating no chemical reactivity at the AuZn/TiN interface.

3.2. ZrB$_2$ thin films

Dependencies of resistivity and stress of ZrB$_2$ films on the dc power and on the argon pressure are shown in Fig. 2. With an increase of power from 100 to 200 W the resistivity changes slowly from 195 μΩ cm down to 175 μΩ cm, but compressive stress increases from 1.0 GPa up to 1.55 GPa. An increase in argon pressure results a small increase in the resistivity and a decrease in stress from -1.7 to -0.38 GPa. X-ray diffraction measurements of ZrB$_2$ films grown at following conditions: $P = 200$ W, $p = 5$ μbar, indicate that ZrB$_2$ has textured polycrystalline structure (hexagonal structure with lattice constants: $a = 0.3196$ nm, $c = 0.3530 \pm 0.0005$ nm). XPS study of the film shown Zr rich compound (Zr 58%, B 42%) with an amount of pure Zr (Zr 3d5/2 peak at 175 eV).

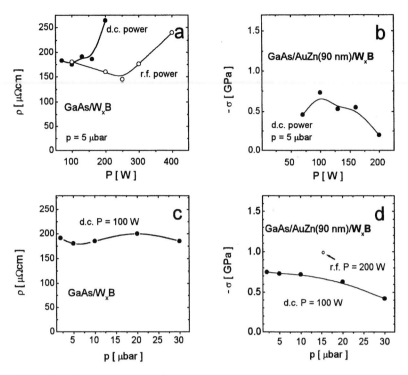

Fig. 3. The dependencies of W_2B film resistivity, ρ, and stress, σ, on the sputtering parameters: power (a and b) and argon pressure (c and d). The results concern the film at the thickness of 100 nm.

When ZrB_2 film is formed at higher argon pressure of 20 μbar, than it is amorphous.

3.3. W_2B thin films

The dependencies of W_2B film resistivity and stress on the rf or dc power, bias voltage and argon pressure are shown in Fig. 3. A minimum of 140 μΩ cm resistivity at rf power of 250 W, zero bias and argon pressure of 5 μbar is observed. When bias voltage of −60 V is applied the resistivity of W_2B film grown at rf power of 200 W and at pressure of 5 μbar decreases to 125 μΩ cm. It has been the lowest value from many our W_2B samples prepared at others conditions. The film grown at above conditions on GaAs/AuZn(90 nm) substrate is under high compressive stress of −1.1 GPa. XRD measurement of that film revealed an coexistence of β-WB and W_2B phases. W_xB layers made at rf power of 200 W, zero bias voltage and at higher pressure of 12 μbar, as well as at dc power in range of 100–200 W are amorphous. The W_xB films grown at dc power are higher electrically resistant and lower stressed than those W_2B films grown at rf power. Two

kinds of W_xB films were chosen for study barrier properties (Table 1).

3.4. Barrier properties

The suppression of As loss by BF during heat treatment can be evidenced comparing As losses from contact structures with contacts without capping layer. During ohmic contact formation of GaAs/AuZn (300 nm) at 420°C for 3 min the As loss has a value of $1.1 \cdot 10^{17}$ atom/cm^2 [3]. One hundred-nm thick TiN film on the contact suppresses As loss at these heating conditions to the level of $1.2 \cdot 10^{16}$ atom/cm^2. Under optimum conditions for AuZn thickness of 90 nm and for TiN growth (350 W, −60 V, 5 μbar, $N_2/Ar = 0.18$) during ohmic contact formation at 320°C for 3 min no As loss has been detected. In the case of AuZn(90 nm)/TiN(75 nm)/Au(100 nm) contact aging at 300°C for 10 h, a small As loss of $0.4 \cdot 10^{15}$ atom/cm^2 was detected. From SEM observation of this contact we concluded that it was due to the microcracks, which take place at microparticles in metallization layer. This effect probably causes a release of stress observed in

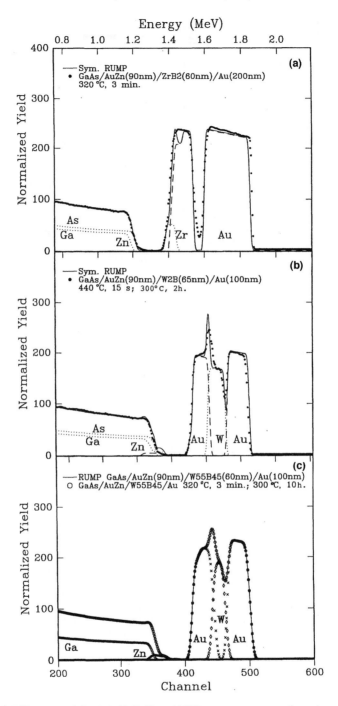

Fig. 4. Two MeV He$^+$ RBS spectra of the GaAs/AuZn(90 nm)/BF/Au contact structure after a heat treatment when: BF is ZrB$_2$(60 nm) film and ohmic contact was formed at 320°C for 3 min. (a), BF is W$_2$B(65 nm) and ohmic contact was formed at 440°C for 15 s and aged at 300°C for 2 h (b), BF is amorphous W$_{55}$B$_{45}$ (60 nm) film and ohmic contact was formed at 320°C for 3 min and aged at 300°C for 10 h.

this layer. After the same heat treatment of AuZn(90 nm)/BF/Au(100 nm), when the barrier film was ZrB_2(60 nm) or W_2B(65 nm) (formed at rf or dc) the As loss was in the level of detection limit of $0.2 \cdot 10^{15}$ atom/cm^2.

Barrier properties of 60-nm tick ZrB_2 and 65-nm thick W_2B (rf) films against to gold diffusion are presented in Fig. 4. When the multilayer contact structure with BF is annealed for ohmic contact formation the changes of RBS spectra are negligible. Only at Ga and As energy edges in RBS spectra small differences have been detected because Zn signal overlaps the edges and it flattens with Zn diffusion through out GaAs/Au/Zn/Au contact. The spectra of the contact structure are stable after heat treatment at 300°C for 2 h. When contact structure with W_xB (dc) film is aged at 300°C for 10 h, the changes in RBS spectra concern the AuZn region, but the interface W_xB/Au is stable.

4. Conclusions

From TiN, ZrB_2 and W_2B films the TiN films reveal the best electrical conductivity. The TiN film deposited at $P = 350$ W, $U_B = -60$ V, $N_2/Ar = 0.18$ and $p = 5$ μbar on the GaAs/AuZn substrate is compressive stressed up to -2.7 GPa and it has low resistivity (45 μΩ cm). The structure and composition of W_xB film depend on the type of sputtering power — high compressive stressed polycrystalline film grows at rf power, less compressive stressed amorphous film grows at dc power. ZrB_2 films have resistivity about 180 μΩ cm and relatively low compressive stress of -0.8 GPa,

which can be reduced at higher pressure during deposition.

All investigated films have properties of diffusion barriers in Au-based metallization to GaAs. They considerably suppress As loss during heat treatment of p-GaAs/AuZn/BF/Au contact structure. Interdiffusion between Au and TiN diffusion barrier after heat treatment at 300°C for 10 h was observed to be very low. No gold penetration was observed by RBS through 100 nm of Au over 60 nm of ZrB_2 or 65 nm of W_2B aged at 300°C for 2 h and over 60 nm of $W_{55}B_{45}$ (amorphous) aged at 300°C for 10 h.

Acknowledgements

This work has been supported by the Polish Committee for Scientific Research trough grant No.8 T11B 05211.

References

[1] Shen TC, Gao GB, Markoc H. J Vac Sci Technol B 1992;10:2113.
[2] Nicolet M-A. Thin Solid Films 1978;52:415.
[3] Piotrowska A, Kaminska E, Guziewicz M, Adamczewska J, Kwiatkowski S, Turos A. Mat Res Symp Proc 1993;300:219.
[4] Haynes TE, Chu WK, Aselage TL, Picraux ST. Appl Phys Lett 1986;49:666.
[5] Chourasia AR, Chopra DR. Thin Solid Films 1995;266:298.

PERGAMON

Solid-State Electronics 43 (1999) 1063–1068

SOLID-STATE ELECTRONICS

TEM studies of the microstructure evolution in plasma treated CVD TiN thin films used as diffusion barriers

S. Ikeda [a,1], J. Palleau [b], J. Torres [b], B. Chenevier [a,*], N. Bourhila [a], R. Madar [a]

[a]*Laboratoire des Matériaux et du Génie Physique, UMR CNRS 5628, ENSPG, BP46, 38402 Saint-Martin d'Hères, France*
[b]*GRESSI CNET/FT, 28 Chemin du Vieux Chêne, B.P. 98, Meylan, France*

Received 10 September 1998; received in revised form 11 January 1999; accepted 24 January 1999

Abstract

In semi-conductor technology, TiN thin films elements are used as diffusion barriers between a copper interconnect layer and a silicon oxide dielectric. Plasma treatment application, by modifying the film microstructure, can improve the film barrier properties and its electrical conductivity. But details of the plasma application effect on the film microstructure evolution and correlations of this evolution with the physical properties are still unclear. To clarify the correlations, the microstructure of a series of TiN thin films deposited using an OMCVD (organo-metallic chemical vapor deposition) technique has been analyzed by transmission electron microscopy (TEM). The films were obtained by cycling a basic synthesis sequence including first a limited film growth and then application of a N_2/H_2 gaseous plasma with various powers and duration times. Films are actually stackings of plasma-treated elementary layers. TEM analyses show that films are made of nanocrystallites and that whereas no texture is observed when no plasma is applied, short-time plasma treatment induces a tendency to ⟨100⟩ texture and if treatment is longer, the direction of texture progressively rotates to ⟨110⟩. A tentative interpretation of this texture evolution has been made in terms of nucleation and growth and correlations between this evolution and the effect on the physical properties have been obtained. © 1999 Elsevier Science Ltd. All rights reserved.

1. Introduction

In semiconductor industry, titanium nitride (TiN) layers are commonly used as diffusion barriers to prevent the conducting material (Cu or W) from diffusing into the silicon substrate through SiO_2 layer. Recently, the organo-metallic chemical vapor deposition (OMCVD) technique has been developed for its desirable step coverage. Tetrakisdimethylamino titanium (TDMAT) precursor has been used to obtain low resistivity TiN films [1]. Although TDMAT has been shown to improve the step coverage of the CVD TiN films, this film resistivity was found too high in comparison with the values obtained with PVD TiN films. This is largely due to the presence of a large amount of carbon and a significant oxygen content in CVD films. By applying a reducing gaseous plasma treatment during film deposition it is possible to improve the film resistivity, by removing carbon and mainly a smaller proportion of oxygen. However, over processing rather degrades the diffusion barrier properties [2]. In the present work, cross-sectional TEM observations have been made to investigate the microstructure changes occurring in TiN films when deposition and

* Corresponding author. Tel.: +33-76-82-64-58; fax: +33-76-82-63-94.
E-mail address: bernard.chenevier@inpg.fr (B. Chenevier)
[1] On leave from National Research Institute for Metals, 1-2-1 Sengen, Tsukuba, Ibaraki 305, Japan.

Table 1
Film synthesis characteristics and microstructure parameter values obtained from TEM analysis

Plasma			S.N.[a]	SiO$_2$ thickness (nm)	TiN			
power	N.E.L.[b] thickness (nm)	duration (s)			thickness (nm)	grain size (nm)	columnar tendency	P.O.D.[c]
–	60	–		1700	120	2–3	poor	no
350 W	5	17	12	1700	75	3–5	poor	100
		40	12	0	60	3–10	medium	100
		90	12	0	55	5–15	medium	110
		150	12	0	50	5–15	strong	110$_{[111]}$
		150	12	850	50	5–15	strong	110$_{[111]}$
	10	35	6	1700	75	5	medium	weak 100
		150	6	0	65	5–10	poor	110$_{[100]}$
		300	6	750	80	5–10	poor	110$_{[111]}$
450 W	10	150	6	850	55	10	medium	110

[a] S.N.: sequence number.
[b] N.E.L.: nominal elementary layer.
[c] P.O.D.: preferred orientation direction. Indices in the P.O.D. column are for a minor component.

reducing gaseous plasma treatment are combined. For brevity concerns, results will be only briefly discussed.

From previous reports found in literature, TiN films deposited under these conditions can be basically described as a random distribution of very small grains or as amorphous material. Film structure X-ray diffraction characterizations usually give patterns with a single and very broad line, roughly corresponding to the TiN 200 line [3].

2. Experimental procedure

As details of the fabrication method have been reported elsewhere [2], we will give here only necessary elements to understand the microstructure and physical properties description. TiN films were obtained in an applied materials machine where film partial deposition and N$_2$/H$_2$ ion-plasma treatment can be sequentially combined: repeated fabrication sequences including an elementary TiN layer deposition followed by plasma treatment can be made. In this work, the elementary deposited thickness was 5 or 10 nm and the fabrication sequence was respectively repeated 12 or 6 times to obtain a final nominal film thickness of 60 nm. Films were grown at a growth rate of 2–3 μm h^{-1}, on top of oxidized (001) silicon substrates kept at 400°C. Nontreated samples were also prepared and served as references to determine the microstructure evolution as a function of plasma parameters. These films were made with a single deposition run.

Plasma ions were produced by a radio-frequency generator and then accelerated to the sample substrate via a d.c. voltage applied on the substrate holder. This allows a major fraction of the ion beam to hit perpen-dicularly the substrate. The applied voltage and the current intensity used in the radio-frequency system define a power parameter. In the power range used here, ion-energy values are of a few tens of electronvolts and increases with the power. In the preparation of our series of samples, powers (P) of 350 and 450 W were used for periods of time (t) ranging between 17 and 300 s. Samples can then be referred as $S_{P,t,ne}$ where P is the power, t the plasma duration and ne the nominal elementary deposited thickness. Main characteristics of the investigated samples are reported in Table 1.

Cross-sectional TEM samples were first mechanically polished by using standard techniques to a thickness of about 40 μm and then were ion-milled in a Baltec-010 machine at 7–5 kV, 1.7–1.0 mA. Conventional TEM observations were made using a JEOL 200CX electron microscope and some high-resolution observations, using a JEOL 4000EX with 0.17 nm resolution.

3. Results

A typical set of electron diffraction patterns is given in Fig. 1. From the diffraction rings indexation, TiN film microstructure can be described as an assembly of very small crystallites with NaCl structure type, i.e. fcc symmetry with a unit-cell parameter, $a = 0.42$ nm ($d_{111} = 0.24$ nm, $d_{200} = 0.21$ nm, $d_{220} = 0.15$ nm). Inhomogeneities in the diffracted intensity along the 111, 200 and 220 rings are indications for preferential grain orientation. The relationships between the symmetry observed in the ring intensity distribution and its orientation to the film normal indicate that in the sample series, 200 and 220 are the main film texture

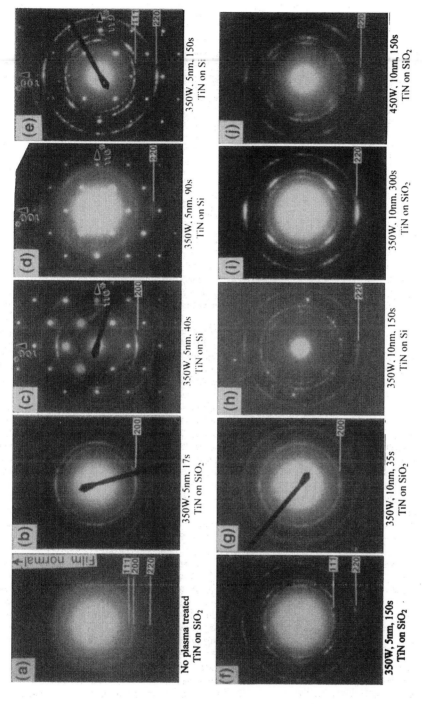

Fig. 1. Selected area cross-sectional diffraction patterns of non-treated TiN film (a) and plasma treated films (b–j). No influence of SiO$_2$ layer on the texture has been detected (see for example (e) and (f)).

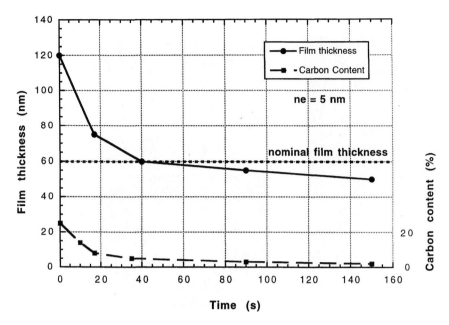

Fig. 2. Comparison of film thickness and carbon content as functions of plasma application duration.

component. Some 111 contribution is also observed. Main TEM results are reported in Table 1.

Cross-sectional TEM observations allowed actual layer thicknesses to be measured and the TiN thickness dependence upon plasma parameter to be determined.

3.1. TiN film obtained without plasma treatment

From plan-view observations made from non-treated films, the microstructure can be described as an assembly of nanonocrystallites with a grain-size distributing between 2 and 3 nm. The intensity of the selected area electron diffraction rings shown in Fig. 1(a) is homogeneously distributed along the ring indicating that no significant texture component is present in this film. The contrast of diffraction rings is faint and the patterns have a high background especially at the inside of 111 ring, suggesting the existence of considerable amount of highly disordered regions. These regions may be made of amorphous material or thick grain boundaries.

3.2. Effect of plasma treatments

Cross-sectional TEM observations of ten films treated with various plasma conditions indicate that in addition to modifications in chemical composition reported earlier, plasma treatment alters the film texture and creates a multilayer-like structure associated with a tendency to a columnar grain organization. Finally, plasma treatment influences the film grain size.

Measured TiN film thickness values (me) are reported in Table 1. For ne = 5 nm, variations of thickness and carbon concentration (from Ref. [2]) as function of plasma duration are compared in Fig. 2. In both curves, a sharp drop is obtained in the first few seconds of treatment. In this range ($t < 15$ s), where carbon concentration is higher than 10%, the film thickness of TEM values exceeds by far the total nominal thickness. For longer time, both curves decrease and saturate (≈ 50 nm and $\approx 2\%$) for $t > 150$ s. Although fewer data are available for ne = 10 nm they exhibit a similar tendency except for very long time plasma applications where a slight increase in film thickness is detected. This last point remains unclear for the present time.

In Fig. 1 Si (110)* planes can be easily recognized. The TiN diffraction rings can thus be orientated in this plane. With increasing parameters a texturing effect starts with a [100] preferential orientation perpendicular to the film surface, for light plasma treatments and for thin elementary layer ($S_{350,17,5}$ for instance). It progressively increases (Fig. 1(c)) with plasma parameters and for heavier treatments and thicker elementary layers, the orientation rotates to [110] ($S_{350,150,5}$; $S_{450,150,10}$). Some 111 texture component is also present for high plasma parameters.

A multilayer-like structure characterized by alternat-

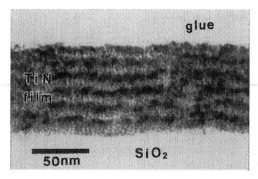

Fig. 3. Cross-sectional view of TiN film plasma treated at 350 W every 10 nm for 35 s.

ing dark and light bands in bright-field cross-sectional views is progressively introduced in the TiN films when plasma parameters increase. This is clearly seen in Fig. 3 ($S_{350,35,10}$). As the light/dark stacking is initiated with a light band at the film/substrate interface, the dark bands can be unambiguously related with plasma affected areas. The dark-band thickness increases with t. In high-resolution obtained from $S_{350,300,10}$, darker bands clearly correspond to larger scale crystallized areas and a close examination of these areas indicates that a majority of observed lattice fringes are parallel to the interfaces and spaced by 0.15 nm, as expected from the diffraction ring analysis of heavily treated samples.

The typical grain-size in the film is a few nanometers. It varies with the plasma intensity between 2 and 3 nm for the non-treated samples and increases up to 15 nm for the highest plasma parameters.

An additional preferential orientation component is superimposed to the one perpendicular to the film surface in heavily treated samples. It is characterized by local orientation correlations between grains of successive layers leading to the formation of columnar organizations with an axis perpendicular to the film surface (see dark-field image of Fig. 4 obtained from

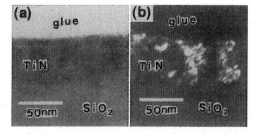

Fig. 4. A cross-sectional view of TiN film plasma treated at 350 W every 5 nm for 150 s. (a) Bright-field image, (b) dark-field image using $g = 220$, normal to the film surface.

$S_{350,150,5}$). The column-diameter is a few tens of nanometers and its length approaches the film thickness. An elementary thickness dependence of the columnar tendency has been found. It is more pronounced for ne = 5 nm.

4. Discussion

In TiN (NaCl structure), the (100) plane has the lowest surface energy since it has the most dense atom arrangement. A nucleation process controlled by surface energy could explain the observed 100 texture for short and low power plasma application. The change to 110 texture can be attributed to an implantation effect of channeling-type [4,5]. In TiN ⟨110⟩ crystal directions are the most 'opened' ones. Ions propagation along these paths will leave crystallites less damaged. Thus alignment of [110] crystallite directions with impinging ion direction will produce a selection of well oriented growth seeds. This will ultimately favor the growth of a majority of crystallites with (110) planes preferably parallel to the film surface.

The time-dependence of the plasma-affected region thickness is closely connected with the evolution of film carbon concentration as a function of plasma duration [2]. Longer plasma exposures have been shown to reduce the amount of carbon [2] and to a minor extent oxygen concentration. Carbon removal occurring in the whole film is probably at the origin of the observed reduction in film thickness. The similar behaviors of carbon concentration and film thickness give some confirmation of this interpretation. Determination by Auger spectroscopy of oxygen concentration profiles has also shown a lower oxygen density limited to plasma affected regions.

The elementary thickness dependence of the film columnar character can be explained in terms of plasma affected-layer inter-coupling. Non-affected thin layers allow affected layers with larger crystallites to couple more efficiently one to the other and thus lead to better columnar shape formation. Therefore a more pronounced columnar character is correlated with higher plasma parameters. Subsequently as higher columnar character very likely favors diffusion along grain boundaries, our observations support earlier reports of poorer barrier properties resulting from heavy plasma treatments applications.

5. Conclusions

TEM observations show TiN films deposited on oxidized silicon substrates and prepared using a combination of OMCVD technique and reducing N_2/H_2 gaseous plasma application are made of very small crystallites with NaCl structure-type. In non-plasma treated films the nanocrystallite size distribution is in the range 2–3 nm and crystallites are randomly oriented. Plasma application introduces a texturing effect beginning with a [100] preferential orientation direction perpendicular to the film surface (for light plasma treatments and for thin elementary layer). It progressively increases with plasma parameters and for heavier treatments and thicker elementary layers, the orientation rotates to [110]. A nucleation process controlled by surface energy could explain the observed 100 texture. The change to 110 texture can be attributed to an implantation effect of channeling-type. A multilayer-like structure is obtained by plasma application. It is described by an alternate stacking of large scale crystallized areas (affected areas) and poorly crystallized (non-affected areas) zones where the grain-size does not exceed 5 nm. TiN film microstructure exhibits a columnar character that increases with the plasma parameters and is likely at the origin of the poorer barrier performance resulting of high plasma parameter application.

Acknowledgements

The authors gratefully acknowledge various stimulating discussions with Dr. Ghetta as well as her critical reading of this paper.

References

[1] Eizenberg M, Littau K, Ghanayem S, Mak A, Yamada Y, Chang M, Shinha AK. Appl Phys Lett 1994;65:2416.

[2] Marcadal C, Richard E, Torres J, Palleau J, Ulmer L, Perroud L. Materials for advanced metallization (MAM '97), European Workshop/Villard de Lans, 16-3-1997. p. 54.

[3] Eizenberg M, Littau K, Ghanayem S, Liao M, Mosely R, Shinha AK. J Vac Sci Technol A 1995;13:590.

[4] Yu L, Harper JME, Cuomo JJ, Smith DA. Appl Phys Lett 1985;47:932.

[5] Yu L, Harper JME, Cuomo JJ, Smith DA. J Vac Sci Technol A 1986;4:443.

PERGAMON

Solid-State Electronics 43 (1999) 1069–1074

SOLID-STATE ELECTRONICS

Texturing, surface energetics and morphology in the C49–C54 transformation of TiSi$_2$

Leo Miglio[a],[*], M. Iannuzzi[a], M.G. Grimaldi[b], F. La Via[c], F. Marabelli[d], S. Bocelli[d], S. Santucci[e], A.R. Phani[e]

[a]INFM Unità di Milano and Dipartimento di Scienza dei Materiali dell'Università, via Emanueli 15, I-20126 Milano, Italy
[b]INFM Unità di Catania and Dipartimento di Fisica dell'Università, Corso Italia 57, I-95121 Catania, Italy
[c]CNR-IMETEM, Statale Primosole 50, I-95121 Catania, Italy
[d]INFM Unità di Pavia and Dipartimento di Fisica "A. Volta" dell'Università, via Bassi 6, I-27100 Pavia, Italy
[e]INFM Unità dell'Aquila and Dipartimento di Fisica dell'Università, via Vetoio 10, I-67010 L'Aquila, Italy

Received 21 September 1998; received in revised form 19 December 1998; accepted 4 January 1999

Abstract

In this paper we analyse the extent of grain orientation before and after the C49–C54 transformation in TiSi$_2$, depending on the substrate microstructure and the lateral dimensions of the film. In the former case, for blanket configuration, we make a comparison to the corresponding evolution in surface roughness, both by AFM and by light scattering measurements. In the second case, an interpretation of the strong texturing occurring in narrow lines, independently of substrate microstructure, is given on the basis of surface energy calculations. © 1999 Elsevier Science Ltd. All rights reserved.

1. Introduction

Despite the recent efforts in understanding the thermodynamic and kinetics aspects of the C49–C54 phase transformation of TiSi$_2$[1], there are several parameters that do influence the actual transformation rate and some of them are difficult to control, such as, for example, the surface cleanness, the purity, and the microstructure of the film. The latter is particularly important in the case of submicron lines, since a strong texturing of the target phase has been reported [2], but it is not known how it depends on the substrate preparation and the parent phase microstructure. It is not

even clear if this effect is produced by thermodynamic or kinetics reasons, nor if the leading role is played by the dimensionality of the line or by the lateral interfaces to the oxide.

The aim of this work is to start casting some light on these problems by considering some crystallographic, morphologic and energetic aspects of the transformation, both in blanket films and in narrow lines. In fact we investigate by XRD measurements the extent of grain orientation before and after the C49–C54 transformation, depending on the substrate microstructure and the lateral dimensions of the film. This is accomplished by a comparison between the data obtained in different experimental configurations, where the contribution from the atomic planes parallel to the sample surface are alternatively enhanced or depressed. In the case of blanket configuration, we make a morphological analysis, both by AFM and light scattering measurements, of the evolution in the surface

---

* Corresponding author. Tel.: +39-02-6448-5217; fax: +39-02 6448-5403.
E-mail address: leo.miglio@mater.unimi.it (L. Miglio)
[1] See for example some relevant contributions in Ref. [1].

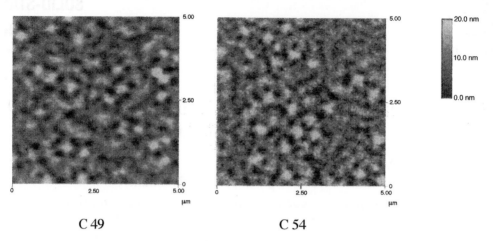

C 49　　　　　　　　　　　　　　　C 54

Fig. 1. Top view by AFM of the initial (C49) and final (C54) surface morphology in the case of phase transformation on a-Si substrate. The larger grain size in the latter case is not visible.

roughness during the transformation, suggesting a connection to the substrate microstructure and to the surface energetics. The latter, as calculated by a semiempirical tight binding molecular dynamics, provides a kinetics interpretation of the strong texturing occurring in submicron lines, in relation to the an estimated hierarchy in external and internal surfaces.

2. Growth and microstructural characterization

Samples were prepared by evaporating Ti on crystalline (100) or amorphous-Si (200 nm thick). The undoped amorphous-Si films were deposited by low-

pressure chemical vapour deposition on thermally oxidized (100) Si substrates, After the amorphous silicon deposition a SiO_2 layer was deposited by plasma enhanced chemical vapour deposition (PECVD) at low temperature and several structures were etched on this layer by a photolithographic process. The latter consisted of several stripes of the same linewidth (1 and 0.5 μm) that were repeated sequentially to form region of several millimeters and a blanket region of several millimeters in size to perform the comparison with the C49–C54 polymorphic transition in not patterned regions. After the pattern definition the wafers were dipped in buffered HF and immediately loaded into an electron beam evaporator for deposition of 17 nm of

Table 1
Surface roughness for blanket films, as measured by AFM and by differential reflectivity

	Anneal. time (s)	σ (nm), AFM 5 × 5 μm	σ (nm), AFM 1 × 1 μm	σ (nm), diff. refl.
Sample series #5	0 (C49)	4.0	3.8	4.5
Substrate: amorphous Si	160	5.5	4.4	8.2
Anneal. temp.: 725°C	240	5.2	5.1	5.8
	320	5.2	4.1	5.5
	500 (C54)	4.6	4.2	3.6
Sample series #4	0 (C49)	11.01	8.24	10.5
Substrate: Si(100)	30	8.92	9.71	9.1
Anneal. temp.: 775°C	50	10.52	9.00	9.8
	70	9.86	9.20	7.3
	90	10.05	8.37	9.6
	110	10.15	7.34	9.3
	150 (C54)	8.40	7.36	8.0

titanium. In some cases the substrate temperature during deposition was 400°C. Annealing at 650°C for 5 min in a nitrogen atmosphere resulted in the formation of a uniform, polycrystalline C49 $TiSi_2$ layer, 40 nm thick, as checked by transmission electron diffraction and X-ray diffraction. In the case of a-Si substrate, a transformation to polysilicon is also found. After etching the unreacted titanium on the SiO_2 regions, isothermal conversion annealing to form the C54 phase was performed by rapid thermal processing in N_2 flux, at different temperatures in the range 750–850°C. The surface morphology was detected by a Digital Instruments Nanoscope III atomic force microscope (AFM) in tapping mode configuration and the vertical roughness σ was derived from the analysis of the AFM images. Fig. 1 displays the initial and final AFM images in the case of a-Si substrate. Despite the much larger size of the C54 grains (some μm) with respect to the C49 ones (decines of nm), the horizontal scale of σ is the same in both cases (i.e. the C49 one). Quantitative values of σ for Si(100) and a-Si are reported in Table 1, along with optical measurements.

3. Optical characterization

The vertical roughness of the film surfaces has been estimated by the amount of diffused light in the visible and ultraviolet spectral range. Measurements in the spectral range 2000–250 nm have been performed on a grating double beam spectrometer (mod. Varian Cary 5) equipped with an integrating sphere.

The diffuse radiation spectra can be analysed through a simple model [3], relating the measured total diffuse reflectance to the vertical root mean square roughness σ (assumed much smaller than the wavelength of radiation). We underline that σ, as evaluated from this model, is in general a function of λ, since the sensitivity of radiation to the surface morphology depends on its wavelength. Therefore we performed our analysis in a restricted range of the spectra, where σ could be expected to be constant (below 600 nm), and where the experimental curves clearly appear to be in agreement with the model. This choice of small wavelengths implies that we are sensitive to the roughness at a small scale (typically, we can appreciate the C49 grains).

The resulting values for σ are reported in Table 1, as compared to AFM measurements. We observe that there is a fairly good agreement in values and an excellent accordance in trend during the phase transformation.

4. Results on surface roughness

The samples deposited on a-Si show a marked increase in σ during the first steps of the transformation, followed by a smoothing of the small scale roughness as the film approaches the C54 phase. The samples deposited on crystalline silicon have generally a larger value of roughness and do not exhibit a clear evolution, apart from a slight decrease in σ from the C49 to the C54 phase.

A comparable optical characterization of σ has been reported by Lavoie et al. [4]. They found the same difference between the films on crystalline and polycrystalline substrates, and almost the same values of σ, as estimated by AFM.

They distinguished between small scale (below 0.5 μm) and large scale (about 5 μm) roughness, on the basis of the scattering angle of the monochromatic light beam. It is interesting to make some remarks.

1. The results of Ref. [4] for the 5 μm scattering on samples deposited on polycrystalline Si exhibit, at the C49–C54 transition, a similar roughness increase as in our case, whereas the 0.5 μm scattering shows a monotonic behaviour more similar to our films deposited on Si(100).
2. The grain size in our C49 films is much lower (20–30 nm) than the one in the samples used in Ref. [4] and there is a clear difference in the texturing occurring on crystalline or polycrystalline substrate.
3. The σ values obtained by optics tend to be closer to the AFM results obtained on a 5×5 μm area than to the ones for the 1×1 μm area. This is related to the fact that optical results average on a much larger area, but can also indicate that our analysis is sensitive to a larger lateral scale than expected.

5. X-ray diffraction characterization

X-ray diffraction measurements on the films have been carried out on a Siemens D5000 diffractometer with Cu K_α ($\lambda = 0.15406$ nm) using the Seeman-Bohlin (grazing angle (G-A)) and Bragg–Brentano (B–B) configuration. For both the measurements we have used a X-ray source voltage of 40 kV and 40 mA of current. In the case of G-A method, we have used the following configuration: X-ray source at grazing angle of 0.7° with respect to the sample, Soller slits, monochromator and detector. Whereas in the case of B–B measurement we have used X-ray source, sample, 6 mm/Ni filter/0.2 mm slits and detector. Nickel and 0.2 mm slits are used to absorb K_β radiation and to improve the resolution, respectively.

Table 2
XRD measurements, both in Bragg–Brentano and grazing angle configurations, as compared to standard powder intensities

Sample	1° step	2° step	Microstr.	Bragg Pl.	B–B	G-A	Powder
#4 Si(100) + Ti 400°C	700°C 40 s	–	no	C49(131)	48%	100%	100%
				C49(200)	100%	–	35%
#5 a-Si + Ti 400°C	700°C 40 s	–	no	C49(131)	100%	100%	100%
				C49(200)	33%	27%	35%
#4 Si(100) + Ti 400°C	700°C 40 s	775°C 150 s	no	C54(311)	100%	100%	100%
				C54(040)	33%	26%	43%
				C54(022)		19%	70%
				C54(331)		35%	45%
#5 a-Si + Ti 400°C	700°C 40 s	725°C 500 s	no	C54(311)	100%	100%	100%
				C54(040)		–	43%
				C54(022)		13%	70%
				C54(331)		11%	45%
Si(100) + Ti RT	700°C 50 s	–	yes	C49(131)	100%	100%	100%
				C49(200)		26%	35%
Si(100) + Ti RT	700°C 50 s	850°C	yes	C54(311)	100%	100%	100%
				C54(040)	100%	28%	43%
				C54(331)	–	17%	45%
a-Si + Ti RT	650°C 300 s	–	yes	C49(131)	100%	100%	100%
				C49(200)	71%	–	35%
a-Si + Ti RT	650°C 300 s	850°C 60 s	yes	C54(311)	9%	50%	100%
				C54(040)	100%	100%	43%
a-Si + Ti 400°C	650°C 300 s		yes	C49(131)	100%	100%	100%
				C49(200)			35%
a-Si + Ti 400°C	650°C 300 s	850°C 60 s	yes	C54(311)	100%	100%	100%
				C54(040)	100%	–	43%
				C54(022)	–	40%	70%

The reasons why we compared the G-A and B–B configurations are the following.

1. Grazing incidence onto the side of the sample pointing towards the X-ray tube means that the X-ray beam does not penetrate deeply into the sample and that only deflection information from the sample surface is obtained. In this way the unfavorable ratio between the signal from the covering layer or sample surface and the signal from the substrate obtained with B–B arrangement is greatly improved.
2. The texture of thin films usually shows rotational symmetry with any (*hkl*) plane oriented parallel to the substrate surface. Accordingly, in the symmetrical B–B arrangement, reflection from the (*hkl*) plane is enhanced, a fact which greatly facilitates detection of crystalline material or profile analysis of (*hkl*) planes. On the other hand, if reflections from planes slanted with respect to (*hkl*) planes are of interest the G-A diffraction has an indisputable advantage, but the (*hkl*) contributions are quenched. Therefore, a comparison between the reflections appearing in both the diffractometer configurations gives an easy and impressive way to establish if the films are grown in a preferred crys-

tallographic direction respect to the substrate plane.

6. Results of XRD measurements

XRD analysis in B–B as well as in G-A configuration of the as-grown C49 $TiSi_2$ films showed two peaks, corresponding to (131) and (200) diffractions, in both blanket and patterned films, depending on the preparation procedure. The former is the most intense peak observed in powder diffraction and always appears, both in B–B an G-A configurations. On the other hand, the (200) peak is quenched in some G-A measurements and is correspondingly enhanced in the B–B data, both in the Si(100) blanket case and a-Si, RT grown lines. In such cases a clear indication of some texturing occurs. For what concerns C54 films, we note a strong tendency to (040) texturing in the patterned case, as can deduced from the intensity ratio $I(311)/I(040)$ in B–B configuration. In polycrystalline material it is expected to be 2.3, as can be estimated from powder diffraction. For blanket C54 samples we measured a value very close to that one, whilst it

Table 3
Surface tensions, as calculated by tight binding molecular dynamics in the relaxed configurations

	Atoms per simulation cell	Initial slab (eV/atom)	Relaxed slab (eV/atom)	Surface tension (eV/A^2)
C49(131)	216	−7.50	−7.61	0.08
C49(100)	216	−7.50	−7.59	0.14
C49(010)	216	−7.63	−7.65	0.12
C49(001)	216	−7.60	−7.65	0.10
C54(311)	216	−7.69	−7.69	0.11
C54(100)	288	−7.73	−7.73	0.11
C54(010)	288	−7.68	−7.69	0.14
C54(001)	192	−7.72	−7.72	0.12

reduced to the range between 0.1 and 1 in patterned samples. A further test of texturing comes from G-A analysis: in this configuration the planes which are nearly parallel to the sample surface are not seen in the diffraction pattern. We found that if $I(311)/I(040)$ < 1 the (040) diffraction peak is missing in glancing angle XRD, but for the case of a-Si + Ti at RT, which is, however, a situation where some texturing of the C49 parent phase has already occurred.

Our data (Table 2) are in agreement to those reported in Ref. [5], where texturing of C54 TiSi$_2$ grown on polycrystalline Si is not observed in blanket films but occurs in patterned sample. Moreover, from the variation of the $I(311)/I(040)$ ratio, they found an increase in texturing by decreasing the linewidth, with values of the ratio between 0.04 and 1 for lines as large as 1 μm. This is in quantitative agreement with our results, in spite of the different procedure in sample preparation. Moreover, it should be noted that this is the first time that texturing has been reported for patterned films grown on (100) silicon substrates.

7. Tight binding estimation of surface energies

The interatomic potential that we used in our simulations is made out of two parts: one is the cohesive contribution provided by the summation of the occupied electronic states, as obtained by a tight binding semiempirical scheme fitted to ab initio calculations of C54 TiSi$_2$, the other one is the repulsive contribution, as given by the summation over first and second neighbours of a short range two-body potential, fitted to structural and elastic experimental data of C54 and C49 phases. The forces acting the on the atoms are calculated as derivatives of the interatomic potential, where the attractive part is obtained in a Hellman–Feynman scheme. The molecular dynamics code that we used to evaluate the atomic displacements allows for variations in volume and shape of the simulation cell (which contains hundreds of atoms) and the tem-

perature is kept constant by heat exchanges with an external thermostat. After checking that the atomic position for the bulk, periodic phases do correspond to an equilibrium situation, we evaluated the surface energies by relaxing the atomic positions at 300 K, for several slabs with different surface orientations. In particular, the oriented simulation cells have been built with vacuum interfaces along one direction and periodic boundary conditions along the remaining two. Sufficiently large simulation cells have been adopted, so that the interactions between the two surfaces of the slabs are avoided and possible surface reconstructions are allowed. After some picosecond relaxation, the surface tension is calculated by comparing the total energy at 300 K with that one obtained for bulk simulations at the same temperature. We evaluated the surface energies for the low index planes of C54 and C49 TiSi$_2$, and a couple of orientations which give rise to most common diffraction peaks. In Table 3 we report our results, indicating also the size of the simulation cell, the energy per atom in the slab with ideal atomic positions and that one for the final configuration in our simulation, where some surface relaxation or reconstruction has occurred. It is clear that C49(100) and C54(010) are the most energetic surfaces, both before and after relaxation.

8. Discussion

The first comment relating our XRD and tight binding results is that the surfaces which happen to be upward oriented, i.e. C49(100) and C54(010), are the most energetic ones. This can be understood by considering that both nucleation and growth of internal surfaces in a columnar model are easier for the other, less energetic ones. This advantageous configuration is likely to overcome the other ones especially if the kinetics is slower, as in the case of low temperatures and of few nucleation sites. In addition to that, from Table 2 it appears that when a preferential orientation of

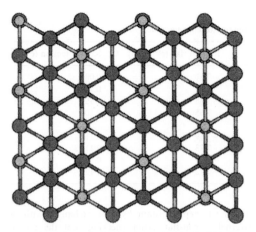

Fig. 2. Simulation cell of the C54, as taken by the [001] direction, with (010) at the top and (001) by the side, periodically repeated in our calculation. The Si atoms, large circles, lay in the first (001) plane, the Ti atoms, small light circles, are in the second plane.

C49(100) is present a misoriented (or less oriented) growth of C54 is obtained. For blanket films, samples #4 and #5, the roughness is larger in the latter case, indicating that for oriented C49 the fewer microstructural degrees of freedom which were present in the initial Si(100) configuration, and the larger surface energy obtained at the end, do generate a large surface roughening. The latter is nearly maintained after the phase transition to C54, as the massive transformation seems not to modify the morphology of the surface, despite the large structural reorientation of the bulk.

In the case of submicron lines, we confirm that a strong tendency to C54(010) oriented growth is found for C49 misoriented parent phase. The fact that additional texturing of the line along the C54(100) direction is found in literature [2] can be guessed by Fig. 2, where we see that this atomic plane (top view) can contain just Si atoms (large circles). In fact, the latter is most likely to match the Si or SiO_2 by Si rebonding at the interface, generating a lower interface tension than for the other faces (side views in Fig. 2). Thus, oriented growth in this configuration should be kinetically favoured.

Acknowledgements

Work partially supported by CNR grants #CT88.00042.PF30 and #97.01355.PS48.

References

[1] Tung RT, Maex K, Pellegrini PW, Allen LH, editors. Silicide thin films: fabrication, properties and applications. MRS Symp Proc 1996;402.

[2] Harper JME, Rodbell KP. J Vac Sci Technol B 1997;15:763.

[3] Bennett HE, Portens JO. J Opt Soc Am 1961;51:123.

[4] Lavoie C, Martel R, Cabral C Jr., Clevenger LA, Harper JME. In: Cammarata RC, Chason EH, Einstein TL, Williams ED, editors. Structure and evolution of surfaces, MRS Symp. Proc 1997;440:389.

[5] Roy RA, Clevenger LA, Cabral Jr. C, Saenger KL, Brauer S, Jordan-Sweet J, Bucchignano J, Stephensen GB, Morales G, Ludwig Jr. KF. Appl Phys Lett 1995;66:1732.

PERGAMON

Solid-State Electronics 43 (1999) 1075–1078

SOLID-STATE ELECTRONICS

Sputtered tungsten films on polyimide, an application for X-ray masks

J. Ligot*, S. Benayoun, J.J. Hantzpergue, J.C. Remy

Laboratoire de Physico-Chimie des Surfaces ENSAM-CER d'Angers, 2 Bd du Ronceray, BP 3525, 49035 Angers, France

Received 9 October 1998; accepted 21 January 1999

Abstract

Among processes of lithography for submicron technologies, X-ray lithography shows the great advantage of high resolution. In the fabrication of X-ray masks, tungsten (W) is one of the alternative absorber materials that can be used instead of gold, but it requires a low stress film with high density. Tungsten films were deposited on Kapton and Upilex polyimides using a triode discharge system. The dependence of thin W films properties on the working argon pressure is reported. X-ray diffraction, AFM and curvature radius measurements were used to characterize the coatings in terms of microstructure and mechanical properties. Average stress and density were found to be significantly dependent on the argon pressure. Microstructural analysis indicated that the grain size and the proportion of metastable phase, β-W, could vary in a wide range. The adhesion of W to polyimide verified with the scratch testing technique seemed very acceptable. © 1999 Elsevier Science Ltd. All rights reserved.

1. Introduction

Since the early 1980s X-ray lithography (XRL) has been developed as an alternative technique to ion-beam and e-beam lithography. XRL allows both higher resolution and a faster process. Among the main components of an XRL system, resist, mask, aligner and source, the mask requires a high-accuracy process [1]. X-ray mask consists of a pattern defined by an absorber heavy metal (100–600 nm) [2] deposited on a X-ray transparent membrane. Gold has widely been used as an absorber material for first X-ray masks [3]. At the present time, the great experience of refractory materials induces the integrated-circuits manufacturers to switch from gold to tantalum or tungsten (W). W is an alternative to gold because of

its similar X-ray absorption coefficient in the wavelength range (0.7–1.2 nm) useful for XRL.

For the thin membrane (1–3 μm), several materials have been used such as Si, SiC, Si_3N_4, BN and polyimide. Polyimides, in regard to their chemical and thermal properties, satisfy the required conditions for materials allotted for submicron technology such as the elaboration of X-ray masks. In this paper two polyimides are studied: Kapton (PMDA-ODA) and Upilex S (BPDA-ODA). Their thermal expansion coefficients are respectively 20×10^{-6} K^{-1} and 8×10^{-6} K^{-1}. Considering its thermal expansion coefficient, W $(4.3 \times 10^{-6}$ $K^{-1})$ appears to be a more attractive absorber material than gold $(16.2 \times 10^{-6}$ $K^{-1})$ for a deposition on Upilex S membrane.

However, a great stopping power demands a high density absorber material. Moreover, the elimination of distortion in X-ray masks requires an absorber film with low average stress and uniformity of the stress distribution [2].

---

* Corresponding author. Fax: +33-02-41207387.
E-mail address: jerome.ligot@angers.ensam.fr (J. Ligot)

0038-1101/99/$ - see front matter © 1999 Elsevier Science Ltd. All rights reserved.
PII: S0038-1101(99)00027-1

2. Experimental

2.1. Sample preparation and deposition system

Substrates for film deposition included both Kapton and Upilex S polyimides. Before introduction into the sputter chamber, an organic wet cleaning (ethanol and deionized water) was carried out in an ultrasonic bath, followed by nitrogen drying. Lastly, in order to evaporate alcohol and water absorbed by polymers, samples were annealed at 200°C for 2 h.

W films were deposited using a triode discharge system (Trion 400 Alcatel) with pure argon as the sputtering gas. The main advantage of the triode discharge lies in the possibility to independently control sputtering parameters such as argon pressure, target current and accelerating voltage for the ions Ar^+. The filament, the positively biased anode and the magnetic field of the coil generate a plasma actually independent of the target and substrate voltages.

The cylindrical chamber is of 360 mm diameter and 310 mm length. This deposition chamber includes two water-cooled target holders (target diameter = 100 mm). The distance between the substrate holder and the W target (purity: 99.96%) was 120 mm. The substrate temperature was estimated during the sputtering process by using a thermocouple in contact to the substrate holder.

2.2. Deposition process and sample analysis

The base pressure in the sputtering chamber was better than 2×10^{-4} Pa for all the depositions. W films were deposited for different argon pressures: 0.2, 0.7 and 1.2 Pa. The target voltage was fixed at −1100 V and the target current varied from 75 to 115 mA according to the working pressures. Prior to any deposition, the target surface was sputter-cleaned for 30 min while the substrates were protected by a screen. The substrate temperature was estimated to be lower than 55°C.

The thickness of the W layers was determined by Talystep profilometer measurements achieved on Si wafers. The crystalline structure of these films was determined by means of X-ray diffraction technique. Their surface morphology was examined using atomic force microscopy (AFM). AFM imaging was carried out in air in the contact mode with a commercial instrument (Autoprobe CP, Park Scientific Instruments). Experiments were conducted with a 2 μm cantilever and a monocrystalline silicium tip (Ultralever UL020, PSI) with a contact force of 10 nN. Rutherford backscattering spectroscopy (RBS) allowed the evaluation of the argon concentration. Average stress of the W films was estimated from the change in the curvature of the polyimide substrate by profilo-

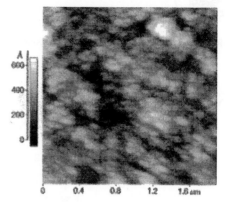

Fig. 1. AFM image of W film of 400 nm thickness at argon pressure 1.2 Pa.

metric measurements (Talysurf). Adhesion between W layers and polyimide was verified with a MST-CSEMEX scratch-tester fitted out with a Rockwell diamond indenter with radius 200 μm.

3. Results and discussion

3.1. Adhesion

Scratch test is a method to estimate the adhesion of films deposited on hard or soft substrates [4]. In order to be sure of the quality of adhesion between W film and polyimide substrate, scratch tests have been achieved for W/Upilex S samples with a load applied on the scratching tip increasing from 0 to 30 N. The observation of these samples has indicated that cracks occurred ahead of the indenter [5]. This damage is typical of hard coatings on softer substrates. However, delamination of the film by adhesion breaking at the interface has never been observed for the differently used deposition conditions. These first results have therefore disclosed a strong adhesion between W as metal and Upilex S as polymer.

3.2. Microstructure

Sputtered films microstructure is a strong function of most prominent deposition parameters such as substrate temperature and working pressure [6]. At a low substrate temperature, less than 55°C in this study, the mobility of the adatoms is low. Moreover, the higher the argon pressure, the lower the energy of incident atoms as a consequence of scattering in the plasma. These collisions randomize the angle of incidence on the growth surface. Both mechanisms result in a shadowing effect during growth and promote a columnar

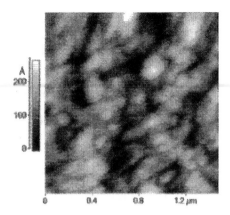

Fig. 2. AFM image of W film of 400 nm thickness at argon pressure 0.2 Pa.

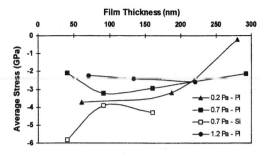

Fig. 3. Average stress versus thickness of tungsten films deposited at different pressures.

porous structure. Fig. 1 depicts the surface topography of a W film grown at higher pressure (1.2 Pa). This surface exhibits the top of typical fine columns, resulting from shadowing effect, with a grain size close to 100 nm.

On the other hand, in Fig. 2, the grains of W film sputtered at 0.2 Pa appears to be larger with an average size of 200 nm. At low pressure, W atoms and neutral Ar particles reflected on the target impinge the film with higher energies. Considering that Ar atoms need an energy higher than a threshold energy to be trapped in the growing film [7], amount of Ar trapped in tungsten increases with decreasing pressure. RBS analyses have indicated Ar contents of 3.1 and 5.3% for respective pressures 1.2 and 0.2 Pa. Thus, at low pressure, three phenomena occur: sputtered atoms impinge onto growing film with a high energy, deposited atoms are heavily peened by the reflected Ar particles, a high amount of argon is trapped into the film. These phenomena seem to hinder the diffusion of W atoms into equilibrium sites on film and favour the formation of metastable β-W phase [8].

The densities of α-W and β-W (W$_3$O) phases are respectively 19.3 and 14.6 g cm^{-3}. The density, calculated with an accuracy of 1 g cm^{-3}, is indicated in

Table 1 as a function of argon pressure for different film thicknesses. Considering that W films grown at pressure 0.2 Pa are less porous, the density obtained at this pressure is the result from a high amount of β phase. X-ray diffraction has revealed a β-W proportion about 45% for a film thickness of 400 nm deposited at 0.2 Pa. At low pressure, the energy of W atoms and Ar particles impinging on the surface of polymer can be sufficient to cause oxygen desorption and chemical damages. Thus, for pressure 0.2 Pa, the additional amount of oxygen present in the W film during the initial stages of growth would promote a higher amount of β phase in comparison with higher pressures. Moreover, during film growth the high energy of W and Ar particles seem to be more favourable for the growth of β-W to the detriment of α-W phase. Consequently, at pressure 0.2 Pa, the density is reduced as thickness increases.

At a pressure of 0.7 Pa, the density close to 17.5 g cm^{-3} for different thicknesses is explained by superposing intermediate porosity and intermediate amount of β phase. X-ray diffraction has indicated a β-W proportion about 34% for a film thickness of 400 nm deposited at 0.7 Pa. For pressure 1.2 Pa, the density close to 17.5 g cm^{-3} is attributed to the porous columnar structure and the amount of β phase. Interactions between energetic particles (W, Ar) and polyimide substrate significantly decrease for this two higher deposition pressures. Film structures are dependent on the deposition conditions and less dependent on the chemistry of the substrate and interface.

3.3. Average stress

The average stresses are shown in Fig. 3 as a function of thickness for W films deposited at different pressures on polyimide Upilex S (Pi) and silicon (Si). Average stresses are calculated with an accuracy of 0.5 GPa. In the range of thickness from 50 to 200 nm, the average stress remain almost stable at any pressure.

Table 1
Density of tungsten films versus argon pressure

Ar pressure (Pa)	Film thickness (nm)			
	200	400	600	800
0.2	17.6	16.2	14.4	–
0.7	–	17.7	17.5	17.2
1.2	17.1	17.7	–	17.9

Moreover, the lower the pressure, the higher the compressive stress becomes, as expected for sputtered α-W films. The compressive stresses are generated by the high energies of W and Ar particles according to the "peening effect" model [9]. However, the high amount of β-W observed at pressure 0.2 Pa for a film thickness higher than 200 nm seems to induce the marked decrease of compressive stress. Moreover, the compressive stresses in W films deposited on polyimide are lower than stresses in W films deposited on silicon. This feature would be the result from the weaker tungsten–oxygen interactions during the film growth. Consequently, the amount of β-W in the film is lower and the compressive stresses are increased.

4. Conclusion

This study has pointed out the importance of the argon pressure in the microstructure and the stress of thin W films deposited on polyimide substrates by sputtering. At a pressure of 0.2 Pa, the amount of β phase increases with increasing thickness. Thus, the density decreases as thickness increases. The presence of β-W is assumed to reduce the compressive stress. At higher pressure (1.2 Pa), a porous columnar structure is observed, and both density and compressive stress remain almost stable versus thickness. From this preliminary work, the compactness and the low compressive stress of W films deposited at pressure 0.2 Pa are more suitable for an X-ray mask application.

Acknowledgements

The authors are grateful to Patrick Saulnier (IBT, Angers) for the AFM characterizations and Wilfrid Seiler (LM3, ENSAM Paris) for the X-diffraction analyses.

References

[1] Silverman JP. J Vac Sci Technol B 1997;15:2117.
[2] Iba Y, Kumasaka F, Aoyama H, Taguchi T. J Vac Sci Technol B 1997;15:2483.
[3] Karnezos M, Ruby R, Heflinger B, Nakano H, Jones R. J Vac Sci Technol B 1987;5:283.
[4] Steinmann PA, Hintermann HE. J Vac Sci Technol A 1989;7:2267.
[5] Ligot J, Le Pourhiet S, Leroux P, Benayoun S. Proc. Graines d'Adhesion, Mulhouse, France, 1996.
[6] Thornton JA. J Vac Sci Technol 1974;11:666.
[7] Meyer F, Bouchier D, Stambouli V, Pellet C, Schwebel C, Gautherin G. Appl Surf Sci 1989;38:286.
[8] Kao AS, Hwang C, Novotny VJ, Deline VR, Gorman GL. J Vac Sci Technol A 1989;7:2966.
[9] d'Heurle FM, Harper JME. Thin Solid Films 1989;171:81.

PERGAMON

Solid-State Electronics 43 (1999) 1079–1083

SOLID-STATE ELECTRONICS

Polymer issues in nanoimprinting technique

Frank Gottschalch[a], Thomas Hoffmann[a,*], Clivia M. Sotomayor Torres[a], Hubert Schulz[b], Hella-Christin Scheer[b]

[a]*Institute of Materials Science and Department of Electrical Engineering, University of Wuppertal, Gauss-Str. 20, 42097 Wuppertal, Germany*
[b]*Micropatterning in Electrical Engineering, University of Wuppertal, Gauss-Str. 20, 42097 Wuppertal, Germany*

Received 2 July 1998; received in revised form 15 December 1998; accepted 16 January 1999

Abstract

We have studied the suitability of poly(methylmethacrylate) (PMMA) of different molecular weight for large area pattern transfer by embossing structures above the glass transition temperature (T_g) of the specific polymer. The molecular weight covers a range of one order of magnitude ($M_w \sim 5.0 \times 10^4$–8.1×10^5). This range was chosen in order to obtain information regarding the flow properties that we expect to depend strongly on the molecular weight at a specific temperature. Large area pattern transfer were tested by applying a stamp with both densely packed and isolated features. The feature size ranged from 100 μm down to 450 nm. At a processing temperature of 90°C above T_g we found clear indications that flow is sufficient to transfer large, isolated features even into the polymer with the highest M_w. Problems of incomplete material transport can be related to local inhomogeneities of the imprint due to a lack of parallelism between the stamp and the sample. At 50°C above T_g incomplete flow effects were observed over the whole area and for all molecular weights. This was observed only with large, isolated structures whereas small, periodic features showed a well defined transfer. © 1999 Elsevier Science Ltd. All rights reserved.

1. Introduction

A major problem in the further development of nanotechnology is the lack of a low-cost, high throughput manufacturing technology suitable for the fabrication of sub-100 nm structures. The need for such a technology emerges clearly from the SIA roadmap that predicts 100 nm gate lengths from the year 2007 onward. It is expected that optical lithography will be limited to approximately 100 nm. Other techniques such as X-ray and ion beam lithography are rather expensive and are proven only on a laboratory scale. At the moment the only proven production technique is e-beam lithography which is limited in its throughput due to sequential writing.

Alternative techniques which may have the potential for mass production have been developed in the last years on a laboratory scale [1–11]. They are based on printing or embossing methods where the latter has similarities to injection moulding or moulding as applied to CD production and micromechanics. Embossing methods differ from conventional lithography in the way the pattern transfer is carried out. Instead of the chemical contrast, common to conventional methods, a thickness relief or contrast is created. The relief is obtained by imprinting a rigid mould with the desired pattern into the resist. In a subsequent pro-

* Corresponding author. Tel.: +49-202-439-2232; fax: +49-202-439-3037.
E-mail address: thomasho@uni-wuppertal.de (T. Hoffmann)

Table 1
Overview of most the important parameters of the PMMA used in this study

Polymer	M_w[a]	T_g (°C)	Thickness[b] (nm)	Solution (wt%)
Plex 6535-F	$\sim 8.1 \times 10^5$	123	400 ± 50	5.99
Plexidon M 449	$\sim 4.8 \times 10^5$	121	420 ± 50	8.34
LP 8423/64a + b	$\sim 1.8 \times 10^5$	109	390 ± 50	12.25
LP 12383/64b	$\sim 5.0 \times 10^4$	85	390 ± 50	15.03

[a] The molecular weight was estimated from the Staudinger index given by the manufacturer.
[b] The thickness values were obtained after the pre-bake step.

cessing step the thinner parts of the relief have to be etched through to obtain a suitable mask for all subsequent processes.

In the case of polymeric resists the polymer thin film is kept above the glass transition temperature T_g where viscous flow of the polymer may occur in the time scale of the experiment. A sufficient viscous flow of the polymer is necessary to allow for displacement of polymeric material from below the elevated structures of the stamp to regions with recessed stamp features. It is obvious that one of the key issues for well defined pattern transfer from the mould to the polymer is the control of the mass transport [6]. The technique has become known as nanoimprint (NIL) [1–3] or hot embossing lithography (HEL) [5]. Despite rather simple experimental set-ups used so far the technique is capable to resolve pattern sizes down to 10 nm and to fabricate devices, e.g. single electron transistors [2]. In a variation of this technique the pattern is embossed into a viscous liquid of a photopolymerisable monomer and subsequently fixed by irradiation through the UV-transparent mould [9].

We have started to investigate NIL on a laboratory scale with a simple experimental set-up for imprinting areas of 2×2 cm^2 size. Our first results showed that two major problems arise with respect to the pattern transfer [6]. First, the problem of sticking which has been already recognised by various authors with proposed solutions based on release agents [3] or anti-adhesive layer on the stamp [7,8] and, second, the problem of mass transport and viscous flow which has not yet been fully addressed. In much of the previously reported work by other authors periodic patterns of small feature sizes, which are ideal to mass transport, have been used. Due to the balance between elevated and recessed stamp features and their proximity, only a small amount of the polymer has to be displaced over a short distance. On the other hand, isolated, large scale features mark the other extreme for mass transport. In this case a large amount of material has to be displaced over a large distance on a time scale that allows for high throughput.

In this paper, we report on first results concerning large area ($\sim 100 \times 100$ μm^2) pattern transfer of atactic PMMA of different molecular weight.

2. Experimental details

Commercial available PMMA (Röhm) of different molecular weights was spin-coated onto substrates of 2×2 cm^2 cut from a silicon wafer. The polymer was dissolved in methoxy-propyl-acetate (Shipley) in different concentrations. The concentration for a specific polymer solution was chosen in order to obtain similar thickness values of the spin-coated layers. The thickness of the resist was measured after pre-baking the polymer at 200°C for 1 h under atmospheric pressure with a profilometer (DEKTAK II) and was approximately 400 ± 50 nm. Table 1 summarises the most important parameters of the polymers used, such as molecular weight M_w, the glass transition temperature T_g, the thickness after pre-bake and the solution concentration.

After cutting, the silicon substrates were cleaned in trichlorethylene, acetone, iso-propanol and deionised water each for at least 3 min in an ultrasonic bath and blown dry with nitrogen.

We used a single 2×2 cm^2 poly-Si stamp for all experiments with an intrusion depth of about 200 nm. Their patterns were obtained by conventional UV lithography and subsequent dry etching. Before every imprint the stamp was cleaned in acetone, iso-propanol and deionised water each for 3 min in an ultrasonic bath and blown dry with nitrogen.

Our previous study on P(MMA/BMA) copolymer of 4.8×10^5 molecular weight showed good results when imprinted at pressures of 100 bar and at a temperature of 90°C above T_g. Therefore, we decided to investigate the pattern transfer under pressures of 100 bar and at two different temperatures, 50 and 90°C above T_g. In order to keep the stamp and the resist coated substrate parallel an elastic O-ring was used between the sample and the stamp support platform. The pressure taken by the O-ring introduces an uncertainty in the pressure values applied to the resist during imprinting. We esti-

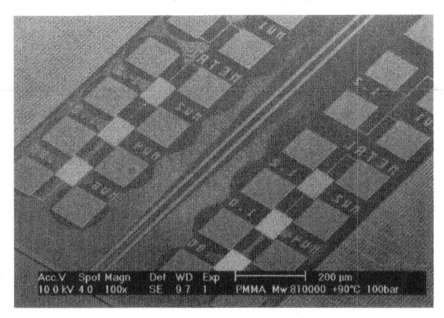

Fig. 1. SEM micrograph of the pattern transferred into the resist ($M_w \sim 8.1 \times 10^5$) after imprinting at $T_g + 90°C$ with 100 bar. Squares of 100×100 μm size correspond to imprinted structures from which the polymer has to be removed during the imprint process. The displaced material appears darker in contrast.

Fig. 2. Local complete material flow between isolated contact pads. The pads are surrounded by a region of displaced polymer that appears darker in contrast. The bright areas that appear in between the pads correspond to periodic structures (same imprint conditions as in Fig. 1).

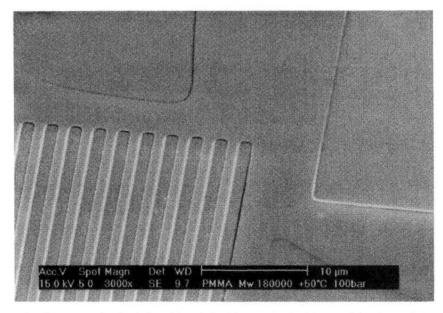

Fig. 3. Comparison between transfer of periodic and large isolated features. On the right part of the micrograph the corner of an imprinted contact pad is visible. The dark border to the left and to the bottom of the pad correspond to the front of displaced material. The imprint was carried out at $T_g + 50°C$ and with a M_w of approximately 1.8×10^5.

mate that the O-ring takes about 25% of total pressure applied, thus the value of 100 bar is overestimated by a similar amount.

In order to imprint, the sample and the stamp were heated to the processing temperature at which the pressure was applied. Immediately after applying pressure the whole set-up was cooled to about 10°C below T_g and the stamp and sample were separated. The pressure was maintained constant during cooling of the set-up. Due to the different times required to cool the various components of the press, the holding times were approximately 10 and 5 min for high and low process temperatures, respectively. After imprinting, an approximately 20 nm thick layer of gold was sputter-deposited on top of the resist and the obtained patterns were inspected by scanning electron microscopy.

3. Results and discussion

A typical result of imprints carried out at $T_g + 90°C$ is depicted in Fig. 1. The SEM micrograph covers an area of about 1.2×0.8 mm^2. Various patterns are transferred into the resist of high molecular weight ($\sim 8.1 \times 10^5$). The central stripe contains imprinted contact pads of 100×100 µm^2 size which appear as squares of brighter contrast. Some of theses squares are surrounded by circle-like structures which appear

darker in contrast (see lower part of Fig. 1). The pads correspond to elevated structures on the stamp, thus material has to be displaced from below these structures during imprinting. The material is first displaced by the pads itself and later flattened by contact of the recessed areas of the stamp with the resist. This leads to a front of displaced material which is elevated with respect to the initial level of the resist after spin-coating. In the upper part of the micrograph no such fronts of displaced material are visible, hence the material has been displaced completely. The homogenous contrast of the areas around the pads is indicative of the absence of topographic features (see Fig. 2 for a more detailed view). The surface appears to be flat due to compression of the resist by the recessed areas of the stamp. Problems of material transport have only been observed with large area, isolated features, e.g. contact pads. However, on the same imprint a large number of contact pads with the quality of pattern transfer as shown in Fig. 2 were observed. The local differences in material flow may be indicative of a lack of parallelism between the stamp and the resist coated substrate. All polymers investigated exhibited the same behaviour independent of molecular weight.

Problems of material transport are much more severe at imprint temperatures of $T_g + 50°C$. They appear on the whole area examined and not only locally, as observed with imprints at $T_g + 90°C$. Fig. 3

shows an example of an imprint carried out with the polymer of low molecular weight ($\sim1.8 \times 10^5$). It should be noted that the material is displaced only close to the border of the pads, whereas the periodic structure did not show such problems in the centre of the periodic pattern. Similar results were obtained for polymers of different molecular weight. The lack of parallelism makes it hard to judge at present if the flow of the polymer depends on molecular weight. No unambiguous proof of a such a trend can be given at the moment.

Sticking has not been a severe problem in this experiment even though no release agent and adhesion promoter were applied. Our previous results have indicated that under optimum conditions ($T_g + 90°C$) sticking becomes less important [6]. In this study sticking was also of little importance even at lower imprint temperatures. This may be due to the different polymer used (PMMA instead of the MMA/BMA copolymer [6]) and/or an additional step with trichlorethylene which has been introduced in the cleaning procedure. As we have used commercial PMMA products the degree of purity of the polymer, in particular, if additional release agents are added to the polymer by the manufacturer, remains unclear.

4. Conclusions

At temperatures sufficiently above T_g, even high molecular weight resists show complete material transport with features that pose challenges to imprinting. The incomplete displacement of the resist appears only locally. It is very likely that these local inhomogeneities are due to lack of parallelism during imprinting, which may be introduced by dust particles, the borders of the stamp and substrate pieces and our rather simple experimental set-up. The lack of parallelism may lead to an inhomogeneous distribution of pressure during imprinting. Besides sticking this appears to be the most severe problem for large area transfer of patterns. Our study confirms previous results which have shown that periodic patterns on a sub-micron scale are significantly easier to imprint even under non optimum conditions.

Acknowledgements

We are grateful to Dr. J. Apohelto for the gift of stamps. We acknowledge the provision of PMMA by Röhm (Germany).

References

[1] Chou SY, Krauss PR, Renstrom PJ. Appl Phys Lett 1995;67:3114.
[2] Chou SY, Krauss PR, Renstrom PJ. J Vac Sci Technol B 1996;14:4129.
[3] Chou SY, Krauss PR, Zhang W, Guo L, Zhuang L. J Vac Sci Technol B 1997;15:2897.
[4] Casey BG, Monaghan W, Wilkinson CDW. Microelectron Eng 1997;35:393.
[5] Jaszewski RW, Schift H, Gobrecht J, Smith P. Microelectron Eng 1998;41–42:575.
[6] Scheer H-C, Schultz H, Hoffmann T, Sotomayor Torres C-M. EIPBN 98 Conf., Chicago. J Vac Sci Technol B 1999;16:3917.
[7] Gröning P, Schneuwly A, Schlapbach L. J Vac Sci Technol A 1996;14:3043.
[8] Jaszewski RW, Schift H, Gröning P, Margaritondo G. Microelectron Eng 1997;35:381.
[9] Haisma J, Verheijen M, van den Heuvel K. J Vac Sci Technol B 1996;14:4124.
[10] Kumar A, Whitesides GM. Appl Phys Lett 1993;63:2002.
[11] Biebuyck HA, Larsen NB, Delamarche E, Michel B. IBM J Res Develop 1997;41:159.

PERGAMON

Solid-State Electronics 43 (1999) 1085–1089

SOLID-STATE ELECTRONICS

Nanometer scale lithography on silicon, titanium and PMMA resist using scanning probe microscopy

Emmanuel Dubois*, Jean-Luc Bubbendorff

IEMN/ISEN, UMRS CNRS 9929, Avenue Poincaré BP 69, 59652 Villeneuve d'Ascq, France

Received 19 June 1998; accepted 4 January 1999

Abstract

Scanning tunneling microscopy (STM) and atomic force microscopy (AFM) nanolithography techniques based on local oxidation of silicon/titanium and electron beam exposure of PMMA are described. It is shown that a 10 nm resolution can routinely be achieved using tapping-mode AFM-based anodization of silicon and titanium operated in air. The thickness and width of oxide stripes are studied as a function of the applied probe-sample voltage and the speed of the tip. Exposure of PMMA resist (950 K, 3%) is also demonstrated using contact-mode AFM to control the tip/surface interaction through a constant force and field emission of electrons to expose the resist. © 1999 Elsevier Science Ltd. All rights reserved.

1. Introduction

Single electronics has recently attracted much attention for future electron device generations and strongly motivates investigations in the field of lithography and pattern transfer in the nanometer regime [1,2]. However, further downscaling of device dimensions rapidly points out the limitations of conventional optical and high energy e-beam techniques. Alternatively, scanning probe microscopy (SPM) has demonstrated capabilities for atomic-level manipulation and also potential for local modification of a variety of surfaces using different approaches: nanometer sized features have been successfully defined using exposure of resists [3], decomposition of organometallic compounds [4] or selective oxidation of hydrogenated silicon [5], amorphous silicon [6] and titanium [7]. In the present paper, the use of several nearfield microscopy techniques (STM, AFM in contact or tapping mode) are discussed for anodic oxidation of silicon and titanium and for

the localized exposure of PMMA to a low-energy electron beam.

2. Nano-oxidation using scanning probe microscopy

2.1. Surface preparation and experimental set-up

The local oxidation of a silicon or metallic substrate is based on an anodization mechanism assisted by the high electric field ($\geq 10^7$ V/cm) between the tip and the sample. The thickness of the oxide layer generated using this technique ranges typically between 1 and 10 nm. As a consequence, special care must be devoted to material preparation for obtaining a clean and flat surface with an RMS roughness below 1 nm. To this aim, silicon surfaces are first thermally oxidized in dry ambient. Subsequent etching in a HF solution leaves a flat surface on which topographic differences due to adjacent (111) oriented planes are observable as shown in Fig. 1. Nanolithography on a metallic surface is performed on a thin titanium layer (5–10 nm) evaporated on the top of an oxidized silicon substrate. All exper-

* Corresponding author.

0038-1101/99/$ - see front matter © 1999 Elsevier Science Ltd. All rights reserved.
PII: S0038-1101(99)00029-5

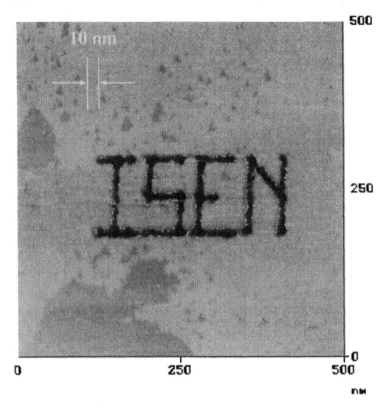

Fig. 1. Nano-oxidation of a (111) oriented silicon surface using STM Bias 3 V, current 20 pA, tip speed 1 μm/s.

iments are performed with a Nanoscope III (Digital Instrument) using a biasing circuit for the control of the anodization mechanism.

2.2. Scanning tunneling microscopy

Fig. 1 shows typical oxide patterns on Si (111) generated under STM operation. After completing the lithography step, the surface imaged in STM mode reveals that an oxide line appears as a depression while the real topography corresponds to a protrusion (volume expansion of SiO_2). This effect may be understood by considering that the STM tunneling current exponentially depends on the distance between the tip and the conducting sample. Oxide lines corresponding to a low surface conductivity causes the tip to move towards the surface giving the pattern an apparent depth. Although, STM lithography and imaging can produce and resolve extremely small topographic features, this operating mode suffers from important limitations as a tool for nanolithography: tip degradation due to reduced conductivity, difficult evaluation of

lithography due to image reversal (apparent depth), slow tip velocity for stable operation in air (Fig. 2).

2.3. Atomic force microscopy

Lithography using atomic force microscopy (both contact and tapping modes) holds two distinct advantages over the STM counterpart. First, the exposure mechanism governed by the electric field can be applied independently of the voltage that controls the tunneling current of the STM. Secondly, the real topography of the oxide patterns is directly given by imaging without needing a conversion from an image that reflects the local density of states as for STM. Fig. 3 shows the linear variation of the oxide height (h) as a function of the applied anodization voltage (V) for silicon substrates with different doping levels. While the occurrence of a threshold voltage has not yet been given a satisfactory interpretation [8], the linear relation found between h and V can be explained through a theory on field-assisted anodization of thin films [9]. The same model explains the logarithmic

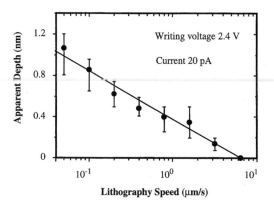

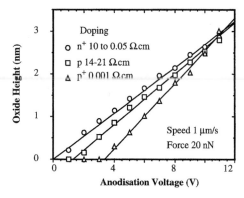

Fig. 2. Apparent depth of oxide stripes performed by STM on Si(100):H, p-doped (B), $\rho = 14$–21 Ω cm at different speeds (reading parameters: voltage -1.5 V, current 20 pA).

Fig. 3. Variations of the oxide height as a function of the anodization voltage Patterns generated by AFM in contact mode on a silicon substrate.

decay of the oxide height when the tip velocity is increased. Fig. 4 illustrates this effect for oxide lines generated in tapping mode on a thin titanium film (9 nm). Beyond the use of the anodic oxide as an efficient masking level for wet or dry etching, the oxide generated by SPM can also serve for the formation of isolating or tunnel barriers. Fig. 5 shows a device fabricated on a thin layer of titanium. Under suitable voltage and speed conditions, the titanium film can be completely consumed by the oxidation reaction, leaving an insulating barrier through which electrons can travel by direct tunnelling or by field-emission, depending on the voltage applied to the source/drain terminals [10]. Fig. 6 corresponds to a typical current–voltage characteristic of the device discussed in Fig. 5. Periodic current variations suggest transport modulated by a mechan-

ism of electron capture and emission by a single trap present in the oxide barrier. This effect is characteristic of random telegraph signal noise (RTS) in small geometry MOSFETs [11,12].

3. Exposure of PMAU resist

Unlike nano-oxidation controlled by the tip-sample electric field, the mechanism of PMMA exposure is controlled by the dose and energy of an electron beam. Due to backscattered electrons, conventional e-beam systems under high energy conditions (20–100 keV) produce exposure over an area much larger than the spot size. The use of low-energy electrons (10–100 eV) emitted from an AFM tip holds the advantage to eliminate proximity effects [13]. The PMMA film was pre-

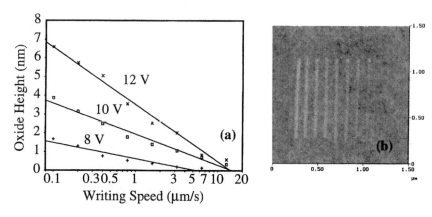

Fig. 4. Nano-oxidation of titanium using TM–AFM. (a) Sensitivity of the oxide height to the tip velocity for different anodization voltages. (b) TM–AFM image of oxide lines with following conditions: (anodization bias 10 V, cantilever amplitude of vibration 4 nm, 9 nm thick titanium film).

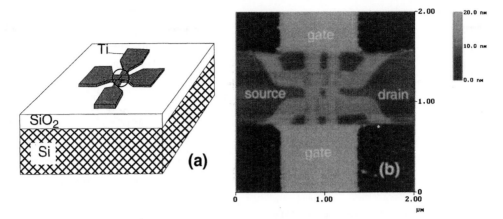

Fig. 5. Direct formation of insulating barriers by oxidation of a thin Ti film (5 nm) (a) layout of a region including contact pads where Ti is lifted-off using optical lithography. TM–AFM based nano-oxidation is used to pattern oxide barriers in the central area (b) corresponding AFM image (anodization voltage 25 V, tip speed 0.25 µm/s).

pared by spinning a PMMA solution 950 K of molecular weight diluted 3% in anisol. The resulting resist thickness is 20–25 nm. The lithography was performed using AFM in contact mode at constant writing speed (0.2 µm/s) and current (2 pA) corresponding to a line dose of 100 nC/cm. The tip voltage varied from −40 to −50 V depending on fluctuations of the resist thickness. The pattern was developed for 120 s in a solution of methyl-isobutyl-ketone (MIBK) and isopropyl alcohol (IPA) in a 1:1 ratio. Fig. 7 shows a scanning electron microscope (SEM) picture of a line with sharp angles for which no widening due to over exposure can be detected. The minimum linewidth proves to be dependent of the resist thickness approximately in a

1:1 ratio, indicating that current spreading takes place in the resist layer under low energy conditions.

4. Conclusion

We have demonstrated that SPM-based oxidation of silicon and titanium offers a high potential for solving problems that conventional lithography can hardly handle. AFM in tapping mode was found to be the best compromise between STM and AFM in contact mode, preserving fine patterns definition while keeping flexibility and eliminating hard tip/sample interactions.

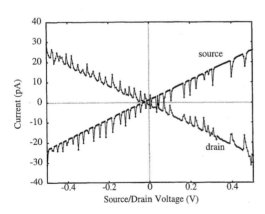

Fig. 6. Current–voltage characteristic of the device shown in Fig. 5.

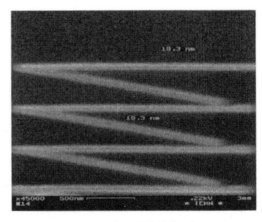

Fig. 7. Pattern generated on PMMA by electron exposure from an AFM tip Thickness of PMMA layer is 20–25 nm, electron dose 100 nC/cm.

AFM in contact mode offers an interesting solution for exposure of thin resist layers at low energy.

References

[1] Ishikuro H, Hiramoto T. Appl Phys Lett 1997;71:3691.
[2] Sakamato T, Kawaura H, Baba T. Appl Phys Lett 1998;72:795.
[3] Soh H, Wilder K, Atalar A, Quate C. In: Symp on VLSI Technology, 1997. p. 129.
[4] McCord MA, Kern DP, Chang T. J Vac Sci Technol B 1988;6:1877.
[5] Dagata JA, Schneir J, Harary H, Evans C, Postek M, Bennett J. Appl Phys Lett 1990;56:2001.
[6] Kramer N, Jorritsma J, Birk H, Schonenberger C. J Vac Sci Technol B 1995;13:805.
[7] Sugimura H, Uchida T, Kitamura N, Masuhara H. Jpn J Appl Phys 1993;32:553.
[8] Ley L, Teuschler T, Mahr K, Miyazaki S, Hundhausen M. J Vac Sci Technol B 1996;14:2845.
[9] Stiévenard D, Fontaine PA, Dubois E. Appl Phys Lett 1997;70:3272.
[10] Matsumoto K, Ishii M, Segawa K. J Vac Sci Technol B 1996;14:1331.
[11] Ohata A, Toriumi A, Iwase M, Natori K. Appl Phys Lett 1990;68:200.
[12] Tsai M, Muto H, Ma TP. Appl Phys Lett 1992;61:1691.
[13] Majumdar A, Oden P, Carrejo JP, Nagahara LA, Graham JJ, Alexander J. Appl Phys Lett 1992;61:2293.

PERGAMON

Solid-State Electronics 43 (1999) 1091–1094

SOLID-STATE
ELECTRONICS

Nanometer patterning of epitaxial CoSi$_2$/Si(100) by local oxidation

Q.T. Zhao*, M. Dolle, L. Kappius, F. Klinkhammer, St. Mesters, S. Mantl

Institut für Schicht- und Ionentechnik, Forschungszentrum Jülich, D-52425 Jülich, Germany

Received 17 October 1998; received in revised form 19 December 1998; accepted 23 January 1999

Abstract

Nanometer patterning of single crystalline CoSi$_2$ layers on Si(100) by local oxidation was studied. Epitaxial CoSi$_2$ layers with thicknesses around 20 nm were grown on Si(100) by molecular beam allotaxy. A nitride layer was deposited on the surface of the silicide and subsequently patterned along the $\langle 110 \rangle$ direction by optical lithography and dry etching. Rapid thermal oxidation was then performed at a temperature of 950°C in dry O$_2$ ambient. During oxidation SiO$_2$ forms on the unprotected regions of the CoSi$_2$ layer. The silicide in this region is pushed into the substrate. Near the edges of the nitride mask the silicide layer thins and finally separates from the protected part. Using this patterning method highly uniform gaps with a width of 70 nm between the two silicide layers have been obtained. It was shown that the separation gap is not only dependent on the oxidation parameters, but also on the thickness and the width of the nitride mask due to the stress effects. Possible applications of this new technique are discussed. © 1999 Elsevier Science Ltd. All rights reserved.

1. Introduction

Silicides are always used as the contact and interconnect materials in modern VLSI and ULSI technologies. Polycrystalline silicides have been used as the source, drain and gate contacts of MOSFETs [1]. Epitaxial silicide, such as CoSi$_2$, has superior properties as compared to the polycrystalline silicide due to the excellent interface sharpness and the lack of grain boundaries. As the feature size of the devices decreases to nanometer, these properties play a more important role. Nanopatterned epitaxial silicides provide a new approach to the future silicon nanoelectronics [2]. Patterning of the silicide nanostructures, however, is

very difficult because of the lack of suitable gases for reactive ion etching. A new method, based on local oxidation of silicide layers, was presented to pattern the epitaxial CoSi$_2$ layers [3,4]. This technique is similar to the well known local oxidation of silicon (LOCOS) technique which is used to form the field oxides for MOSFETs. Standard optical lithography and thermal oxidation are the main processes involved. During oxidation SiO$_2$ forms on the unprotected regions of the CoSi$_2$ layer. The silicide in this region is pushed into the substrate. Near the edges of the mask the silicide layer thins and finally separates from the protected part. Using this method uniform gaps with distance of 200 nm have been obtained by wet oxidation [4].

In this article, we study the patterning of thinner CoSi$_2$ layers ($\sim$20 nm) by rapid thermal oxidation. We will show that minimum feature sizes of less than 50 nm can be achieved.

* Corresponding author. Tel.: +49-2461-613-663; fax: +49-2461-614-673.

E-mail address: q.zhao@fz-juelich.de (Q.T. Zhao)

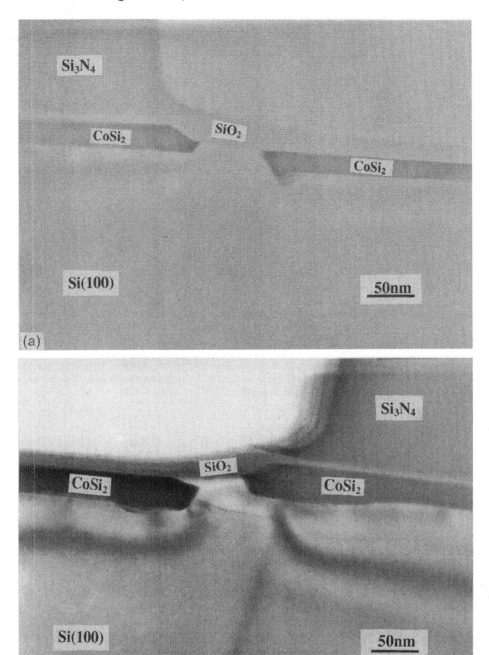

Fig. 1. XTEM micrographs of an epitaxial CoSi$_2$ layer with a thickness of 20 nm patterned by local oxidation at 950°C for 3 min. (a) The nitride mask was 3 μm wide. The gap measures 70 nm. (b) The nitride mask was 2 μm wide resulting in a gap of only 50 nm.

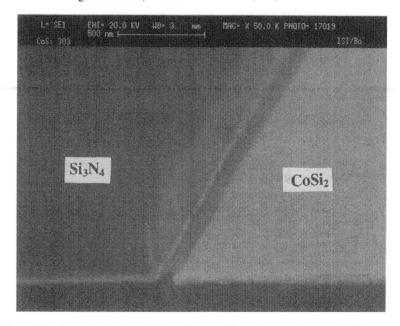

Fig. 2. SEM micrograph showing a highly uniform gap (70 nm) between the separated silicide parts.

2. Experimental

Epitaxial CoSi$_2$ layers with thicknesses around 20 nm were grown on floating zone phosphorous doped Si(100) (> 1 kΩ cm) substrates by molecular beam allotaxy (MBA) [5,6]. An undoped Si epi-layer of 140 nm was grown at a substrate temperature of 600°C. Then the CoSi$_2$ layer was grown by MBA in two steps: First, silicide precipitates were embedded in epitaxial silicon by codeposition of Co and Si at a substrate temperature of 400°C. During deposition, the Si deposition rate was kept constant throughout the whole growth process, the Co deposition rate was ramped up linearly to a maximum concentration of about 31 at% in the middle of the process and then kept constant. Second, subsequent rapid thermal annealing (RTA) at temperature of 1000°C for 40 s was performed to form continuous epitaxial surface CoSi$_2$ layers. A SiO$_2$ layer with a thickness of less than 10 nm was formed on top of the silicide during RTA because of 90%N$_2$ + 10%O$_2$ ambient. This SiO$_2$ layer together with the subsequently deposited Si$_3$N$_4$ layer were used as the oxidation mask for the following local oxidation. MeV He ion channeling measurements have shown that the CoSi$_2$ layers have a high crystalline quality with minimum yields of 4%. Their specific electrical resistivity was measured to be 14 μΩ cm at room temperature.

For local oxidation experiments a Si$_3$N$_4$ oxidation mask with a thickness ranging from 200 to 300 nm was deposited on top by plasma enhanced chemical vapor deposition (PECVD). The mask was patterned by optical lithography and dry etching with CHF$_3$/CF$_4$. Mask lines with widths from 1 to 3 μm were aligned along the ⟨110⟩ directions. Rapid thermal oxidation (RTO) was performed at 950°C in dry oxygen. The resulting structures after oxidation were investigated by scanning electron microscopy (SEM) and cross-section transmission electron microscopy (XTEM).

3. Results and discussion

Fig. 1 shows the XTEM micrographs of an epitaxial CoSi$_2$ layer patterned by local oxidation at 950°C for 3 min with different mask line widths. The nitride mask line widths in Fig. 1(a) and (b) are 3 and 2 μm, respectively. The thickness of the CoSi$_2$ layer is 20 nm before oxidation. In the unprotected regions SiO$_2$ forms on top of CoSi$_2$ layer. The oxide thickness is about 20 nm. The breakdown fields of the SiO$_2$ layers on CoSi$_2$ were measured to be 5–8 MV cm^{-1} approaching nearly the quality of thermal oxide on silicon. During RTO the oxidant (O$_2$) diffuses to the SiO$_2$/CoSi$_2$ interface where two chemical reactions take place: the dissociation of CoSi$_2$ and the oxidation of silicon. The excess Co diffuses through the silicide layer to the lower CoSi$_2$/Si interface to form CoSi$_2$ again, while the excess Si forms SiO$_2$. The latter process gives rise to the growth of the typical SiO$_2$ 'bird's beak' underneath the edge of the nitride mask. 70 and 50 nm gaps between the protected and the unprotected CoSi$_2$ parts formed for the mask line widths of 3 and 2 μm, respectively. The silicide is displaced vertically by only ~10 nm. This is different from the behavior of thick silicide layers, where the silicide in the unprotected regions is pushed into the substrate vertically with a large distance (200 nm) during oxidation [4].

For applications a high lateral uniformity of the gap is necessary. In order to investigate the lateral uniformity of the separation, the oxide was removed by buffered HF etching. The resulting structure is shown in a SEM micrograph of Fig. 2 for a mask line with a width of 3 μm. We can see that the gap of 70 nm is highly uniform. A nitride mask line width of 2 μm results in a narrower gap (Fig. 1(b)) as compared to the 3 μm wide mask, but unfortunately several CoSi$_2$-bridges connecting the two CoSi$_2$ parts were found in the SEM micrographs. For mask lines with a width of 1 μm no separation was found. TEM results showed that near the edges of 1 μm wide nitride mask lines the thickness of CoSi$_2$ layer only decreased (not shown).

The patterning of a 20 nm CoSi$_2$ layer with a nitride mask thickness of only 200 nm was also studied. Our results indicated that for such thinner nitride masks longer RTO is necessary to obtain a uniform separation. A uniform gap of 90 nm was found for 3 μm wide mask line after local oxidation at 950°C for 5 min, but for the narrower mask lines a lot of CoSi$_2$ bridges still existed.

The patterning results show that the separation depends on the thickness and the width of the mask. This is due to the stress distribution generated in the silicide by the nitride mask. We found that the stress generated by the oxidation masks plays a key role in this patterning process. For thicker and wider nitride masks, stress peaks at the edge of the mask is much more pronounced, and thus the separation occurs easier and more uniform. A Monte Carlo simulation, considering the stress dependence of the dissociation of CoSi$_2$ and the oxidation of Si at the SiO$_2$/CoSi$_2$ interface as well as the stress dependent diffusion of Co atoms in the CoSi$_2$ layer and at the CoSi$_2$/Si interface, has been established. It could be shown that the simulation predicts well the evolution of the micrographs during local oxidation [7].

One of the applications of this patterning technique is to fabricate Schottky barrier MOSFETs (SB-MOSFETs) which show some advantages when the feature size reaches nanometer order [2]. The channel region with length less than 100 nm can be defined by our patterning method instead of electron beam lithography. Epitaxial CoSi$_2$ layers have good Schottky contacts with barrier heights of 0.64 eV on n-type Si and 0.46 eV on p-type Si. Our first results on SB-MOSFETs using this patterning method are promising. Further investigations of the devices are still under development.

4. Summary

A nanometer patterning method for thin epitaxial CoSi$_2$ layers based on local oxidation was investigated. Separation of CoSi$_2$ layers takes place near the edge of the nitride mask during rapid thermal oxidation. The separation depends on both the mask thickness and the mask width because of the stress distribution generated by the oxidation mask. Highly uniform gaps of 70 nm between the protected and unprotected CoSi$_2$ layers were fabricated. The vertical displacement of the oxidized part amounts only ~10 nm. This technique can be used to fabricate ultrashort channel SB-MOSFETs.

Acknowledgement

The author Q.T. Zhao would like to thank the Humboldt Foundation of Germany for their support.

References

[1] Maex K. Mat Sci Eng R 1993;11:53.
[2] Tucker JR, Wang C, Shen T-C. Nanotechnology 1996;7:275.
[3] Mantl S, Dolle M, Mesters St, Fichtner PFP, Bay HL. Appl Phys Lett 1995;67:3459.
[4] Klinkhammer F, Dolle M, Kappius L, Mantl S. Microelectr Eng 1997;37/38:515.
[5] Mantl S, Bay HL. Appl Phys Lett 1992;61:267.
[6] Mantl S. J Phys D: Appl Phys 1998;31:1.
[7] Mantl S, Kappius L, Antons A, Löken M, Klinkhammer F, Dolle M, Zhao QT, Mesters St, Buchal Ch, Bay HL, Kabius B, Trinkaus H, Heinig KH. Proc Mat Res Soc 1998. In press.

PERGAMON

Solid-State Electronics 43 (1999) 1095–1099

SOLID-STATE ELECTRONICS

Processing and characterisation of sol–gel deposited Ta_2O_5 and TiO_2–Ta_2O_5 dielectric thin films

A. Cappellani[1], J.L. Keddie, N.P. Barradas*, S.M. Jackson

University of Surrey, Guildford, Surrey GU2 5XH, UK

Received 30 June 1998; received in revised form 19 September 1998; accepted 3 January 1999

Abstract

High-dielectric thin films of Ti-doped Ta_2O_5 were deposited on n^+-type silicon substrate using the spin-on sol–gel process. Doping levels of 8 and 46 TiO_2 mol% were used. Following deposition, films were processed at temperatures between 600 and 900°C using rapid thermal annealing in N_2O. Spectroscopic ellipsometry (SE) and Rutherford backscattering spectrometry (RBS) were used to determine the thickness and the composition of the thin films and the interfacial reaction layers. Metal–insulator–semiconductor capacitor structures were fabricated and impedance–frequency measurements were carried out to measure the dielectric constant of the deposited films. Results from both RBS and SE showed that a SiO_2 layer is formed at the Ta_2O_5/Si interface during processing, but the titanium doping inhibits the kinetics of its formation. We found that the dielectric constant of the highly Ti-doped Ta_2O_5 film was 78% greater than that of Ta_2O_5 sol–gel film processed under similar conditions. © 1999 Elsevier Science Ltd. All rights reserved.

Keywords: High k dielectrics; Thin films; Sol-gel; Ellipsometry; Rutherford backscattering; Simulated annealing

1. Introduction

The dielectric constant of dielectric thin films limits the degree of miniaturisation of capacitive components, which form the basis of many memory devices, such as DRAMs. All DRAM chips manufactured to date use capacitors containing dielectric films of SiO_2 and/or silicon nitride [1,2]. Capacitors dielectrics thinner than those now being used will suffer leakage due to Fowler–Nordheim tunnelling (around below 100 Å). Many compounds with high dielectric constant are being widely investigated, including tantalum pentoxide (Ta_2O_5), which is already accepted in manufacturing facilities due to its compatibility with current microelectronics fabrication procedures. However, its relative dielectric constant [3,4] is at best 35 and a better control has to be obtained over the film composition, properties and the interface between the high-dielectric layer and the electrode [1]. By incorporating 8 mol% TiO_2 (another acceptable manufacturing material) into bulk Ta_2O_5 Cava et al. [5] have succeeded in increasing the dielectric constant to 126.2. The dielectric constant of polycrystalline TiO_2 reaches about 100, but as a result of its large leakage currents [2] it is not attractive as pure material but as a dopant on the basis of its own polarisability.

The bulk material produced by Cava et al. [5] is attractive because of its high dielectric constant. We are not aware, however, of any report of production of

---

* Corresponding author. Tel.: +44-1483-259-827; fax: +44-1483-534-139.

E-mail address: n.barradas@surrey.ac.uk (N.P. Barradas)

[1] Present address: Siemens AG, ZT ME 1, Corporate Technology, Otto-Hahn-Ring 6, 81730 Munich, Germany.

thin films of the high dielectric constant material. Here we report our efforts in achieving the long-term aim of high-dielectric thin films with low dielectric loss. We report the characteristics of thin films of tantalum pentoxide (Ta_2O_5) and titanium-doped tantalum pentoxide (8 and 46 mol% TiO_2) produced by sol–gel deposition. The sol–gel deposition method has been selected because the dopant level can be precisely controlled, it has a relatively low cost, and it can be performed easily onto any substrate [6,7]. Spectroscopic ellipsometry (SE) and Rutherford backscattering (RBS) were used to determine the thickness and the composition of the thin films and of the interfacial reaction layers. Impedance–frequency measurements were carried out to measure the dielectric constant of the deposited films.

2. Experimental details

Both the Ta_2O_5 and Ta_2O_5–TiO_2 films were produced using the sol–gel process in which metal alkoxides are hydrolysed in a solvent in presence of a catalyst. Condensation reactions cause polymerisation to occur and allow inorganic polymers to be deposited from a sol onto many types of substrate. The metal alkoxide precursors used were tantalum ethoxide $[Ta(OC_2H_5)_5]$ and titanium ethoxide $[Ti(OC_2H_5)_4]$, and the acid catalyst was HCl. A sol for Ta_2O_5 has been prepared by mixing $Ta(OC_2H_5)_5$, HCl and H_2O in C_2H_5OH at room temperature under a mild agitation, in the molar ratio 1:7.11:0.051:49.3. For Ta_2O_5–TiO_2 film sol, a mixture of $Ti(OC_2H_5)_4$ and $Ta(OC_2H_5)_5$ in C_2H_5OH was used to produce doping levels of 8 and 46 TiO_2 mol%. The polymerisation reactions were suppressed after 10 min by addition of excess C_2H_5OH to the mixture, stopping gelation until the films were deposited.

The method used to deposit the films was the spin-on process [6,8]. The As^+-doped Si substrate was flooded with the appropriate sol, and was spun at 2000 rpm for 30 s on a Precima spin coater, normally used for photoresist deposition. For most samples, the deposition has been performed by a multiple coating process in order to achieve the required film thickness. After each deposition, the sample was placed on a hot plate to drive off the excess solvent, and to avoid cracking of thick films. After deposition, the films have an amorphous structure, still contain some solvent and are very porous. To produce denser crystalline films, the samples were processed at temperatures between 600 and 900°C using RTA in N_2O ambient. The selected temperature was rapidly reached after a step at 400°C for 1 min, performed to minimise stress in the film. The presence of oxygen is required to achieve a complete oxidation of the organic residues in the cer-

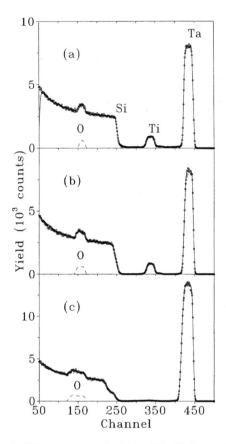

Fig. 1. RBS spectra of samples (a) 2 – TiO_2–Ta_2O_5 annealed at 700°C; (b) 6 – TiO_2–Ta_2O_5 annealed at 900°C; and (c) 5 – Ta_2O_5 annealed at 900°C. The points and the solid lines are the data and the fits respectively. The dashed lines are the partial fitted oxygen spectra. The position of signals corresponding to the different elements present are indicated.

amic structure. The heat treatment densifies the structure, and at sufficiently high temperatures, crystallisation occurs. Standard annealing in air was also performed up to 600°C with a rate of 90°C/min. By encouraging densification prior to crystallisation, RTA is expected to produce denser films [8] and to minimise reaction with the substrate due to the reduced processing time.

The dielectric constant of the samples was determined by fabricating a metal–insulator–semiconductor (MIS) structure. The As^+-doped Si substrate (resistivity 4.2×10^{-4} Ω cm) was used as bottom electrode, while 2 mm aluminium dots were evaporated on the film surface as top electrodes. The impedance of the system was measured over the range 0.5–10 MHz.

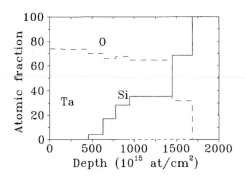

Fig. 2. Depth profile used to fit the RBS data of sample 5 (Ta_2O_5 annealed at 900°C, as shown in Fig. 1c).

All samples were analysed with variable angle spectroscopic ellipsometry (SE), and the data were analysed using the Cauchy model to describe the dispersion of the index of refraction n of the films. For the Ta_2O_5 films, the effective medium approximation (EMA) model was also used, which takes into account the presence of nano-sized air voids ($n = 1$) in the material.

For the RBS measurements a 1.5 MeV He^+ beam at normal incidence was used. The scattering angle was 160°. The combinatorial optimisation simulated annealing (SA) algorithm [9] was used to analyse the data. We have previously shown [10–12] that SA is

well-suited to effectively deconvolute the elements present in different layers. Plural scattering [13] and pile-up [14] corrections were applied to the data. RBS determines the areal density of the different elements, and therefore depth units will be given in at/cm². These can be converted to real thickness if the density of the material is known, or inversely the density can be calculated given the thickness [8].

3. Results and discussion

The RBS spectra obtained for some of the samples are shown in Fig. 1, together with the theoretical spectra obtained from the SA fit. The fits are excellent, and allow us to determine the stoichiometry and thickness of the layers present. No SiO_2 layer can be found in the titanium-doped sample annealed at 700°C (sample 2, Fig. 1(a)), within the sensitivity of the technique (about 100×10^{15} at/cm²). This is not the case for the titanium-doped sample annealed at 900°C (sample 6, Fig. 1(b)): The thicker O signal and the less steep Si edge signal indicate the presence of a thin SiO_2 layer between the Si wafer and the deposited layer, which was confirmed by the fit results. Finally, comparison of Fig. 1(b) and (c) shows that the presence of Ti inhibits the kinetics of the formation of the SiO_2 layer, which is much thicker in the absence of Ti, for the same annealing temperature. However, an oxide layer is still present in the doped samples annealed at high

Table 1
Summary of film thickness, composition and density. Concentrations have been normalised to 100 at% and hence do not take into account the presence of Cl and As

Sample	t_{ell} as-dep. (Å)	RTA anneal temp (°C)	t_{ell} annealed (Å)	t_{ell,SiO_2} annealed (Å)	[Ta] (at%)	[Ti] (at%)	[O] (at%)	t_{RBS} annealed (10^{15} at/cm²)	t_{RBS,SiO_2} annealed (10^{15} at/cm²)	ρ_{layer} (10^{22} at/cm³)	ρ_{SiO_2} (10^{22} at/cm³)
1	1521	700	1076	55	—[b]	0	—[b]	—[b]	—[b]		
2	1314	700	816	30	15.6	16.3	68.1	535	—[a]	6.5	
3	1596	800	954	397	—[b]	0	—[b]	—[b]	—[b]		
4	1357	800	826	30	13.9	15.2	70.9	599	—[a]	7.3	
5	1553	900	928	1538	25.1	0	74.9	688	933	7.4	6.1
6	1326	900	805	537	13.0	13.7	73.3	593	287	7.4	5.3
7	1113	none	—[b]	—[b]	23.2	2.9	73.9	608	—[a]		
8	1118	600	754	42	23.2	3.0	73.8	523	—[a]	6.9	
9	1077	700	764	42	23.7	2.9	73.4	545	—[a]	7.1	
10	1129	800	747	274	22.7	2.8	74.5	535	168	7.2	6.1
11	—[b]	800	826	40	14.1	14.7	71.2	574	—[a]	6.9	
12	—[b]	900	805	537	13.6	13.6	72.8	591	295	7.3	5.5
13	2031	none	—	40	24.2	0	75.8	760	—[a]		
14	—[b]	700	1076	40	28.2	0	71.8	663	—[a]	6.2	
15	—[b]	800	1035	336	27.3	0	72.7	655	227	6.3	6.7

[a] Below sensitivity limit.

[b] Not measured.

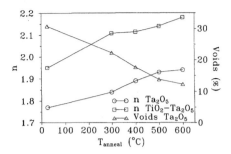

Fig. 3. Refractive index and void content dependence on annealing temperature.

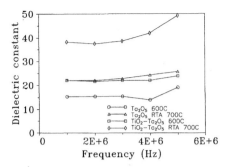

Fig. 4. Real dielectric constant of Ta_2O_5 and 46 mol% TiO_2–Ta_2O_5 films.

temperatures. The depth profile obtained for the undoped sample annealed at 900°C (sample 5, Fig. 1(c)) is shown in Fig. 2, where both the Ta_2O_5 (the stoichiometry determined with RBS was $Ta_2O_{5.7}$, consistent within the fit error) and the SiO_2 layers are clearly seen. The results obtained are included in Table 1. Further, a small Cl contamination (about 5 at% before annealing, reduced to about 2 at% after 900°C RTA) was found in some of the samples, all from the same batch. This probably comes from the HCl catalyst introduced during the sol–gel process. Finally, it should be noted that RBS is not sensitive to chemical binding, only to the amount of each element present, so the results for the Ti-doped samples would be consistent both with the coexistence of segregated Ta_2O_5 and TiO_2 inclusions, and with single-phase Ta_2O_5–TiO_2 films.

The thickness t of the deposited films before and after annealing, and of the SiO_2 layers after annealing, as measured with ellipsometry, are also given in Table 1. As ellipsometry measures real thickness values while RBS determines areal densities, the density ρ of the films can be determined through $\rho(\text{at/cm}^3) = t_{RBS}(\text{at/cm}^2)/t_{ell}(\text{cm})$. The values obtained are given in Table 1. The first observation is that the density values obtained for SiO_2 are on average about 10% lower than the literature crystalline bulk value (6.6×10^{22} at/cm^3), as expected from the amorphous structure of the oxide. The dispersion between values is due to the high error in the RBS determination of the thickness of such thin buried layers.

While the density of the deposited layers is consistently lower than the density of Ta_2O_5 (7.8×10^{22} at/cm^3) and TiO_2 (9.2×10^{22} at/cm^3), it increases with anneal temperature, which shows that densification occurs due to the annealing. Analysis of the ellipsometry results of the Ta_2O_5 films with the EMA model confirms this, as the fraction of voids in the films decreases from 30% in the as-deposited state to 12% after annealing at 600°C, as shown in Fig. 3. However,

some voids remain and the films are polycrystalline since no channelling effect could be observed with RBS. Furthermore, the refractive index of both Ti-doped and undoped films increases with annealing temperature, as expected, reflecting the increased density.

Fig. 4 shows the real dielectric constant values measured for annealed Ta_2O_5 and highly doped Ta_2O_5–TiO_2 films. The values for the doped and undoped as-deposited films were anomalously high (about $\varepsilon_r = 90$). This might be due to the fact that before annealing –OH groups in the amorphous structure are capable of storing charge ($\varepsilon_r = 80$ for H_2O). The real dielectric constant value for Ta_2O_5 films annealed below 600°C, was around 16. The film is still amorphous if annealed at these temperatures, so this value is below the established value $\varepsilon_r = 25$ for Ta_2O_5 [8,15,16]. A mean value $\varepsilon_r = 23$ is reached after annealing at 700°C.

The 46 mol% TiO_2–Ta_2O_5 films show the same trend: the dielectric constant value drops after the first annealing and raises with the annealing temperature. Annealing temperatures below 600°C lead to a dielectric constant mean value of 25. At 700°C annealing, a 78% enhancement of the dielectric constant value to $\varepsilon_r = 41$ is observed, similar to the enhancement found by Cava et al. for similar doping [5]. For higher annealing temperatures the SiO_2 layer formed at the ceramic/Si interface changes the observed dielectric constant of the film, and no conclusions can be drawn about the dielectric constant of the deposited films. The measurement of dielectric constant of the 8 mol% TiO_2–Ta_2O_5 films is in progress.

4. Summary

- Optically transparent, crack-free, thin film forms of Ta_2O_5 and TiO_2–Ta_2O_5 have been deposited on silicon substrate using the sol–gel process with a thickness of about 1600 Å.

- Rutherford backscattering and ellipsometry have been used to characterise the films. The results obtained from the two techniques showed good agreement. The films densify with heat treatment, which is reflected in a reduced fraction of voids and a higher density.

- A SiO_2 interfacial layer is formed at the Ta_2O_5/Si interface during heat treatment at temperatures of 700°C or more. The interfacial layer grows with the annealing temperature. Titanium doping inhibits the kinetics of oxidation, leading to considerably thinner SiO_2 interfacial layers.

- The dielectric constant of TiO_2–Ta_2O_5 thin films was compared against the dielectric properties of Ta_2O_5. Ti-doped films showed a dielectric enhancement of 78% over the undoped film.

- Future investigations will be devoted to the effect of different dopant concentrations on the thin film properties.

References

[1] Kotecki DE. Semiconductor international, November 1996. p. 109.

[2] Streiffer SK, Kingon AI. Nature 1995;377:194.

[3] Ohji Y, Matsui Y. IEDM 1995;111.

[4] Sun SC, Chen TF. IEDM 1994;12.4.

[5] Cava RF, Peck Jr. WF, Krajewski JJ. Nature 1995;377:215.

[6] Brinker CJ, Scherer GW. Sol–gel science. Academic Press, 1990.

[7] Rehg TJ, Ochoa-Tapia JA, Knoesen A, Higgins BG. Appl Optics 1989;28:5215.

[8] Keddie JL, Braun PV, Giannelis EP. J Am Ceram Soc 1994;77:1592.

[9] Aarts E, Korst J. Simulated annealing and Boltzmann machines: a stochastic approach to combinatorial optimization and neural computing. Chichester: Wiley, 1989.

[10] Barradas NP, Jeynes C, Webb RP. Appl Phys Lett 1997;71:291.

[11] Barradas NP, Marriott PK, Jeynes C, Webb RP. Nucl Instr Methods B 1998;136–138:1157.

[12] Barradas NP, Jeynes C, Harry MA. Nucl Instr Methods B 1998;136–138:1163.

[13] Barradas NP, Jeynes C, Jackson SM. Nucl Instr Methods B 1998;136–138:1168.

[14] Jeynes C, Jafri ZH, Webb RP, Kimber AC, Ashwin MJ. Surf Interface Anal 1997;25:254.

[15] Pignolet A, Mohan Rao G, Krupanidhi SB. Thin Solid Films 1995;258:230–5.

[16] Burte EP, Rausch N. J Non-Cryst Solids 1995;187:425.

PERGAMON

Solid-State Electronics 43 (1999) 1101–1106

SOLID-STATE ELECTRONICS

CoSi₂ buffer films on single crystal silicon with Co ions pre-implanted surface layer for YBCO/CoSi₂/Si heterostructures

I. Belousov[a,*], Peter Kus[b], S. Linzen[c], P. Seidel[c]

[a]*Institute for Metal Physics, Ukrainian National Academy of Sciences, Vernadskii Ave. 36, 252680 Kiev, Ukraine*
[b]*Department of Solid State Physics, Comenius University, SK-84215 Bratislava, Slovak Republic*
[c]*Institut für Festkörperphysik, Friedrich-Schiller-Universität Jena, Helmholtzweg 5, D-07743 Jena, Germany*

Received 9 October 1998; received in revised form 14 January 1999; accepted 23 January 1999

Abstract

The Co ions pre-implanting process on Si(100) surface with low dose 10^{15}–10^{16} cm^{-2} at implantation energy 80 keV was used to grow smooth CoSi₂ films as buffer layers for the YBCO/CoSi₂/(Co$^+$)Si structure. The YBCO films have a critical temperature of $T_c^{off} = 86$ K and $T_c^{on} = 80$ K in the YBCO/CoSi₂/(Co$^+$)Si structure. Finally, Co distribution into reacted CoSi₂ layer was established as a result of the two contrary diffusion Co fluxes from Co film into Si bulk and from implanted layers to the Si surface. The ohmic contacts based on CoSi₂/(Co$^+$)Si and YBCO/CoSi₂/(Co$^+$)Si structures have a contact resistivity near 10^{-5} Ω cm² at room temperature. © 1999 Elsevier Science Ltd. All rights reserved.

1. Introduction

The constant interest in cobalt disilicide (CoSi₂) films is caused by the fact, that some of their physical (cubic fluorite structure with a lattice constant smaller by 1.2% than that of Si, chemical stability at high temperature) and electrical properties (low layer and contact resistivity 14 μΩ cm and 10^{-5} Ω cm², accordingly) are the best of known silicides for application as interconnection and contact elements of the silicon microcircuits. Recently was successfully shown an opportunity of application for the CoSi₂ films as a buffer layer between superconducting YBa₂Cu₃O₇₋ₓ (YBCO) film and silicon substrate [1,2]. Such thin film combinations have potential applications in hybrid superconductor–semiconductor technology and promise great profit in microdevices performance [3].

However, one crucial aspect concerning CoSi₂ films is their morphology. The CoSi₂ surface roughness, cracks and pinholes in the silicide buffer are obstacles for YBCO film formation with high quality. It was demonstrated that use of an additional thin Ti or Zr cap layer significantly reduces the CoSi₂ surface roughness and improves the uniformity and the epitaxial quality of CoSi₂ layers, but a ternary Ti(Zr)CoₓSiᵧ layer near the surface is formed in these techniques [4,5]. Moreover, it was shown that the main reasons for surface roughness during silicidation process are local nucleation of silicide phase on crystal defects of the silicon surface and mainly lateral mode growth of the CoSi₂ layer [6].

In this study, low dose Co ion pre-implantation into silicon surface before Co film deposition and silicidation process was carried out to increase surface defects density and to obtain a more homogeneous silicide phase during Co–Si interaction. The planarization

* Corresponding author. +7-44-444-9551; fax: +7-44-444-3432.

E-mail address: belousov@scat.kiev.ua (I. Belousov)

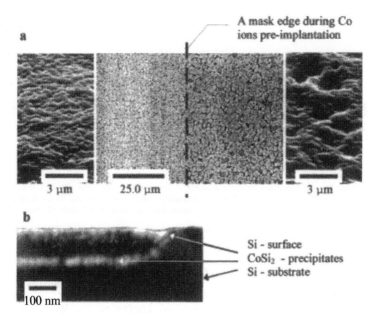

Fig. 1. (a) SEM image of the CoSi$_2$ film surface formed on the pure silicon (right part) and Co ions pre-implanted layer (left part). (b) TEM image of the cross-section for the structure CoSi$_2$/Co$^+$(Si) layer after two steps 650°C, 15 min and 950°C, 3 min vacuum annealing.

phenomena and Co redistribution into reacted Co–Si layer have been investigated as well as electrical properties for YBCO films on the CoSi$_2$ film buffer and CoSi$_2$/(Co$^+$)Si ohmic contact structures.

2. Experimental details

Epitaxial layers of CoSi$_2$ were grown on the Si (100) wafers in an electron-beam oil-free vacuum system with a base pressure of 7×10^{-9} Torr. The YBCO thin films were deposited onto the CoSi$_2$ buffer layer by RF-magnetron sputtering as well as by laser ablation techniques. The details of deposition process, Co, YBCO and CoSi$_2$ films formation are described elsewhere [6,7].

The SCANIBAL SCI 218 implantator was used for the Co ions implantation process at room temperature. The dose implantation was chosen as 10^{15}–10^{16} ion/cm^2 and ions energy 80 keV.

Structural and morphological characterization of the CoSi$_2$ layers, the YBCO/CoSi$_2$/Si structures and buried interface have been performed using transmission electron microscopy (TEM), transmission electron diffraction (TED), scanning electron microscopy (SEM), reflection high-energy electron diffraction (RHEED) and X-ray diffraction (XRD). Chemical composition

was investigated using the Auger electron spectroscopy (JAMP 10s, JEOL).

The electrical properties for the YBCO films and CoSi$_2$/Si ohmic contacts were estimated in four points measurements. The contactless inductive technique allowed us to reveal superconducting properties for discontinuous YBCO films.

3. Results and discussion

The CoSi$_2$ layer, formed with more smooth surface and phenomena of planarization, was observed at interaction Co film and pre-implanted Si surface (Fig. 1). In this case, the right part of the silicon sample was protected by a SiO$_2$ mask during the pre-implantation process; then the protected mask was deleted and CoSi$_2$ film was grown simultaneously on both parts. This fact points to a more homogeneous reaction between Co film and Co-ions implanted Si surface in contrary to pure Si. The cross-section TEM picture for this sample allows to estimate the implanted layer buried under silicide layer as 200 nm (Fig. 1b).

In order to investigate which cobalt distribution into reacted Co/Co$^+$(Si) layers was realized by two steps (650°C, 15 min and 950°C 3 min) vacuum annealing, Co ions pre-implantation onto Si(100) was performed

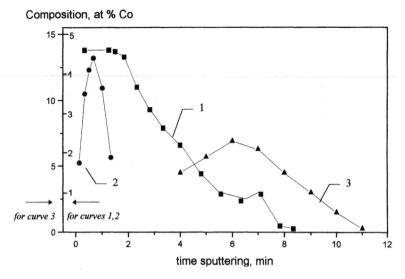

Fig. 2. In-depth Co composition profiles obtained from Auger spectrum for the Co–Si reacted layers after two steps 650°C, 15 min and 950°C, 3 min vacuum annealing (1) 60 nm Co-film/Co$^+$(Si), (2) Co$^+$(Si) and without annealing (3) Co$^+$(Si). Co ions pre-implantation into Si(100): dose 2.0×10^{15} cm^{-2} at 80 keV.

with fixed dose 2.0×10^{15} cm^{-2} at 80 keV acceleration and then 60 nm Co film was deposited (Fig. 2, curve 1). For comparison, a reference sample was annealed at the same temperature regime without Co film (Fig.

2, curve 2). The initial Co ions distribution after implantation process was tested, too (Fig. 2, curve 3).

Fig. 2 shows that vacuum annealing leads to the shift of the Co distribution curve from the initial pos-

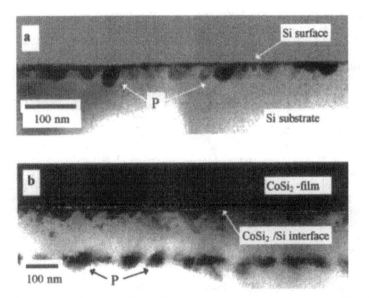

Fig. 3. TEM Image of the cross-section for the (a) Si(100) with Co ions pre-implanted layer (dose 2.0×10^{15} cm^{-2}, $U = 80$ keV) and (b) 60 nm Co-film/Co$^+$(Si) reacted layer after two steps 650°C, 15 min and 950°C, 3 min vacuum annealing. P-arrows point to CoSi$_2$-precipitates.

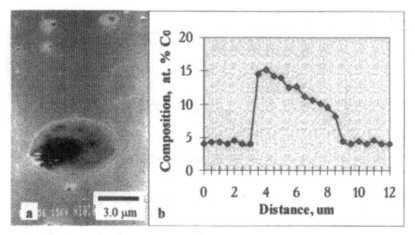

Fig. 4. (a) SEM image of the silicon surface with $CoSi_2$-precipitates (plan view, 60° to surface normal) (b) Co distribution on the Si surface in the Co-precipitate area.

ition (curve 3) to that near the Si surface (curve 2) and a maximum of the Co composition achieved the one like in $CoSi_2$ reacted layer. Thus, the implanted Co ions adsorption at the Si surface results from vacuum annealing. It is possible to assume, that two diffusion Co fluxes from the Co film into silicon bulk and the Co ions from implanted layer to the silicon surface result in cobalt distribution for reacted $Co/Co^+(Si)$

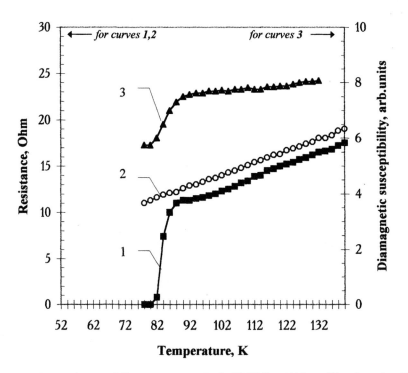

Fig. 5. Resistance and diamagnetic susceptibility vs. temperature for the YBCO films (thickness 100 nm) growing: (1) on the $CoSi_2$ buffer formed on the Co ion pre-implanted Si top; (2) (3) on the $CoSi_2$ buffer formed on the pure Si top.

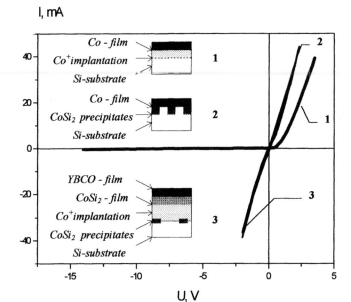

Fig. 6. The current–voltage characteristics for the (1) contact of the metal Co film to Si with Co ions pre-implanted surface layer; (2) $CoSi_2/(Co^+)Si$ and (3) $YBCO/CoSi_2/(Co^+)Si$ structures.

layers. This observation is in a good agreement with the TEM investigation presented in this study. Fig. 3 shows the TEM cross-section pictures corresponding to the Co distribution curves. It is clearly shown that $CoSi_2$ precipitates are formed near the Si surface (Fig. 3a) in the case described by curve 2 (Fig. 2), as well as at the boundary of the Co ions damaged Si layer (Fig. 3b) that is reflected in the case of curve 1 (Fig. 2).

The surface Co distribution was studied by Auger electron spectroscopy with a 0.1-μm probe diameter to clarify the nature of Co-precipitates for Co implanted samples annealed without Co film; the result is shown in Fig. 4. Co-precipitates are disk-shaped particles with different orientation and a diameter up to 5.0 μm. The composition maximum of Co is 15–16 at% and close to the Co composition in the reacted $CoSi_2$ layer.

The surface roughness of the $CoSi_2$ buffer layer results in a change of the electrical properties of the YBCO films in $YBCO/CoSi_2/Si$ structures. Fig. 5 shows the dependence of resistivity on temperature ($R(T)$) for YBCO films (thickness 100 nm) deposited on Si with two types of $CoSi_2$ buffer. The critical temperature of $T_c^{off} = 86$ K and $T_c^{on} = 80$ K was measured for YBCO film deposited on the smooth $CoSi_2$ layer formed on Si with Co pre-implanted surface (Fig. 5, curve 1). The typical $R(T)$ curve as for normal metal without any transition into superconducting condition

was observed for $YBCO/CoSi_2/Si$ structure with a high roughness $CoSi_2$ buffer formed on the pure Si surface (Fig. 5, curve 2). On the other hand, the measurement of the superconducting properties of such films by contactless inductive technique (χ-diamagnetic susceptibility) demonstrates the transition into a superconducting condition within the 86–80 K temperature range (Fig. 5, curve 3). It means that the YBCO film has cracks between superconducting islands and that $CoSi_2$ surface roughness is one reason for it.

The current–voltage characteristics and contact resistance for the $CoSi_2/(Co^+)Si$ and $YBCO/CoSi_2/(Co^+)Si$ structures were measured because ohmic contact between superconducting and semiconducting films is an interesting component for the hybrid devices. In Fig. 6 is demonstrated ohmic-like behavior for current–voltage characteristics of the $CoSi_2/(Co^+)Si$ and $YBCO/CoSi_2/(Co^+)Si$ structures (curves 2 and 3) at room temperature. It should be mentioned that contact of the metal Co film to Si with Co ions pre-implanted surface layer has a diode-like current–voltage characteristic (Fig. 6, curve 1). The contact resistivity for the $CoSi_2/(Co^+)$n-type Si(100) structure was estimated from four point measurements as 10^{-5} Ω cm^2.

4. Conclusion

The Co ions preimplantation process for Si(100) surface with a low dose 10^{15}–10^{16} cm^{-2} at implantation energy 80 keV allowed to grow smooth CoSi$_2$ films as buffer layers for the YBCO/CoSi$_2$/(Co$^+$)Si structure. The YBCO films has a critical temperature of $T_c^{\text{off}} = 86$ K and $T_c^{\text{on}} = 80$ K in the YBCO/CoSi$_2$/(Co$^+$)Si structure. Finally, Co distribution into reacted CoSi$_2$ layer was established as a result of the two contrary diffusion Co fluxes from Co film into Si bulk and from implanted layers onto Si surface. The ohmic contacts based on CoSi$_2$/(Co$^+$)Si and YBCO/CoSi$_2$/(Co$^+$)Si structures have a contact resistivity near 10^{-5} Ω cm^2 at room temperature.

Acknowledgements

This research has been supported by the German BMBF contract No. 13 N 6808. The authors wish to thank all the coworkers for their help, especially V. Svetchnikov for providing us with TEM cross-section micrographs and I. Kossko for Auger electron spectroscopy measurement.

References

[1] Kumar A, Narayan J. Appl Phys Lett 1991;59:1785.
[2] Belousov I, Rudenko E, Linzen S, Seidel P. J Low Temp Phys 1997;106:433.
[3] Qiao J, Yang CY. Mater Sci Eng R Rep 1995;R14:157.
[4] Tung RT, Schrey F. Appl Phys Lett 1995;67:2164.
[5] Naidoo R, Pretorius R, Saris F. Supercond Sci Technol 1995;8:25.
[6] Belousov I, Rudenko E, Linzen S, Seidel P. Microelectron Eng 1997;37/38:581.
[7] Linzen S, Schmidl F, Schmauder T, Schneidewind H, Seidel P, Köhler T. In: Vincenzini P, editor. Superconductivity and superconducting materials technologies. Faenza, Italy: Techa, 1995. p. 273.

PERGAMON

Solid-State Electronics 43 (1999) 1107–1111

SOLID-STATE ELECTRONICS

Excimer lamp-induced decomposition of platinum acetylacetonate films for electroless copper plating

Jun-Ying Zhang*, Ian W. Boyd

Department of Electronic and Electrical Engineering, University College London, Torrington Place, London WC1E 7JE, UK

Received 1 September 1998; received in revised form 15 January 1999; accepted 27 January 1999

Abstract

Photo-induced decomposition of platinum acetylacetonate films using an excimer VUV source of 172 nm radiation is reported. VUV irradiation of a substrate coated with platinum acetylacetonate film results in the formation of platinum, which acts as an activator for copper plating by means of a subsequent electroless bath process. Selective copper patterns can thus be formed by employing patterned VUV irradiation through a quartz contact mask to dimensions approaching the submicron level. The platinum films formed and the platinum acetylacetonate layers used were characterised using ultraviolet spectrophotometry (UV) and Fourier transform infrared (FTIR) spectrometry. The photo-decomposition of the platinum compounds was found to be a function of both the UV dose and chamber pressure during irradiation. The morphology of the copper layers was investigated with a scanning electron microscope (SEM). © 1999 Elsevier Science Ltd. All rights reserved.

1. Introduction

As microelectronic technology advances with device features in the submicron region, the requirements for metallization are becoming increasingly stricter in terms of performance and reliability. Present advanced generation products typically use aluminium based alloys for multi-level wiring. However, for the development of faster devices, the resistance–capacitance (RC) delay must be reduced. Advanced metallization schemes using multilevel metallization (MLM) structures involving Cu are currently being considered since they offer the advantages of low resistivity (1.67 $\mu\Omega$ cm), high electro- and stress-migration resistance and high melting point compared to aluminium based alloys [1–5]. For copper to be fully incorporated, how-

ever, several problems need to be addressed that include rapid diffusion of Cu atoms into silicon and its dielectrics, while its reaction with most metals at temperatures above 200°C must be minimised [3]. Therefore, there is a need for lower temperature deposition technology for Cu deposition applied to microelectronic applications.

Low temperature electroless copper deposition, which provides a highly selective and low thermal budget metallization process, is an attractive, alternative, and low cost technique for the fabrication of advanced copper metallization and interconnect systems for ultra-large scale integrated (ULSI) applications [6–11]. In this technique, photo-induced decomposition of metalorganic layers selectively activates non-catalytic substrate surfaces prior to a subsequent electroless metal plating step. This copper deposition technique offers many advantages over standard thin film deposition methods. The process is low temperature, needs no complicated vacuum equipment, is selective, has fewer process steps, and is relatively simple.

* Corresponding author. Tel.: +44-171-419-3196; Fax: +44-171-387-4350.

E-mail address: jzhang@ee.ucl.ac.uk (J.Y. Zhang)

Previously, we have demonstrated that palladium acetate films can be decomposed with ultraviolet (UV) radiation using different wavelength excimer lamps and used for subsequent electroless metal plating, thereby providing a technology potentially enabling large area and low cost selective electroless metallization [7–11].

In this paper, photo-induced decomposition of platinum acetylacetonate films on different substrates prior to a subsequent electroless metal plating step using excimer UV sources is reported. A selective copper pattern can thus be formed in a simple process using a contact mask. The plating process has a high deposition rate and metal patterns several microns thick can readily be fabricated in this way.

This platinum acetylacetonate, similar to palladium acetate, is an ideal metalorganic precursor for photo-induced metallization since it has high optical absorptivity to allow decomposition at low temperatures which prevents substrate damage, has weak bonding but is sufficiently stable at room temperature, has high solubility in a desired solvent or solvent mixture and easily produces homogeneous films.

2. Experimental

Platinum acetylacetonate (Ptacac) was dissolved in chloroform using concentrations ranging from 0.01 to 0.15 mol/l. The Ptacac films were prepared from solution by the spin-on technique on quartz (Spectrosil B, thickness: 1 mm), Al_2O_3, AlN, crystalline (100) Si and PMMA (polymethylmethacrylate) substrates. UV transmittance spectra were obtained on a UV/VIS spectrophotometer (Perkin Elmer, Lambda14).

The layers were irradiated using an excimer source at a wavelength of 172 nm [8,9] inducing photo-dissociation of platinum acetylacetonate and the deposition of platinum in the irradiated areas. After irradiation, the sample was rinsed in an organic solvent to remove the unexposed platinum acetylacetonate and then immersed in an electroless plating solution at 25°C, where a copper layer was further deposited on the nucleated areas.

For patterning, a contact quartz mask was placed on top of the Ptacac-coated substrate. The thickness of the initial Ptacac films and the Pt layers formed on the quartz at different exposure times was measured by UV spectrophotometry in the 200–400 nm range with an initial precalibration measurement being taken using a ellipsometer (Rudolph AutoEL II). A Fourier transform infrared (FTIR) spectrometer (Paragon 1000, Perkin Elmer) was employed to examine the dependency of the film composition as a function of different irradiation parameters as well as the chemical changes of the C=O, C–C, CH_2, and CH_3 group in the platinum acetylacetonate layers. The morphology of

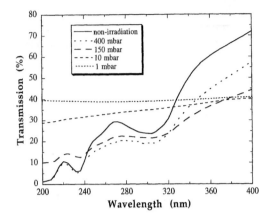

Fig. 1. UV transmission of Ptacac films irradiated at different pressures for a fixed exposure time of 6 min.

the copper layers formed after electroless plating was investigated using a scanning electron microscope (SEM, JSM-T220 A). A standard scotch tape test was used for a qualitative estimation of the adhesion strength of the copper films on the various substrates.

3. Results and discussion

The influence of the chamber pressure on the decomposition rate of Ptacac was studied and Fig. 1 shows the UV transmission of the Ptacac films irradiated at different pressures for a fixed exposure time of 6 min. At high pressures, the UV transmission

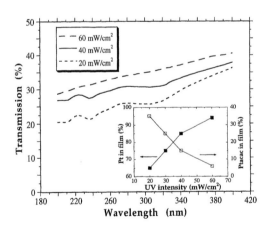

Fig. 2. UV transmission of Ptacac films irradiated under different UV intensities at a fixed pressure of 1 mbar. The inset shows the percentage of Pt and Ptacac in the films as a function of UV intensity.

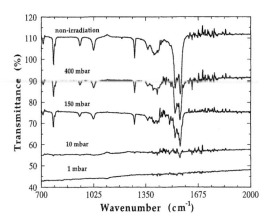

Fig. 3. FTIR spectra of the Ptacac films irradiated under different pressures at a fixed exposure time of 6 min.

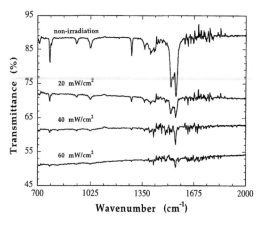

Fig. 4. Evolution of FTIR spectra of the Ptacac films irradiated under different UV intensities at a fixed pressure of 1 mbar.

change was notably less than for lower pressures with the spectra at 400 mbar showing very little change from the unirradiated spectrum at wavelengths below 240 nm. At a pressure of 1 mbar, an almost flat transmission curve is observed. This is typical of what is expected for a pure Pt film [12] and indicates that the Ptacac has almost completely decomposed. Clearly the decomposition rate of Ptacac is more rapid at lower pressures. Two effects will strongly influence this. Firstly the volatile decomposition products from the reaction zone will be easily removed at lower pressures. Secondly at higher pressures, the short wavelength radiation produced by the lamps will be more strongly attenuated by the decomposition productions of the exposed Ptacac. Therefore, higher decomposition rates can be expected at lower pressures.

The influence of the UV intensity on the decomposition rate of Ptacac was studied and the change in the UV transmission of the Ptacac films irradiated at different UV intensities for a fixed two minute exposure time at 1 mbar is shown in Fig. 2. As can be seen, the decomposition rate increases with UV intensity and higher photon fluxes from the excimer lamp lead to more reactivity. The thickness of Pt formed after irradiation of the Ptacac was obtained using the Beer–Lambert Law applied to the change in UV transmission [9]. As the intensity varied from 20 to 60 mW/cm^2, the percentage of platinum formed from the Ptacac changed from 65% to 94% as shown in the inset of Fig. 2. A similar effect has been observed with palladium acetate when irradiated at the same wavelength [9].

Fig. 3 shows the evolution of the FTIR spectra of the films after exposure to the 172 nm radiation for various pressures. The band between 1510 and 1600 cm^{-1} is associated with the C=O stretching vibrations

of the β-diketone ($-[CH_3COCHCOCH_3]$) groups, while the band at 1387–1420 cm^{-1} corresponds to the C–H bending vibration of the alkane ($-CH_2-$, $-CH_3$) groups, and that at 1280 cm^{-1} results from the C–C bond stretching, while the 775 cm^{-1} absorption corresponds to the rocking mode of the CH$_2$ and CH$_3$ groups. All absorption bands decreased with decreasing pressure, practically disappearing at a pressure of 1 mbar after irradiation for 6 min, indicating the decomposition of the acetylacetonate group. At higher pressures the decrease was much smaller than for lower pressures as expected, in agreement with the results of the UV measurements (Fig. 1).

The FTIR spectra of the films also changed with UV intensity. Fig. 4 shows the evolution of the FTIR spectra of the Ptacac films irradiated under different UV intensities at a fixed pressure of 1 mbar for 2 min. All the absorption bands mentioned above decreased with UV intensity as expected.

In the electroless plating process both anodic oxidation of a reducing agent and cathodic reduction of metal ions take place on the irradiated surface [13]. The bath used consisted of a solution of Cu ions, a complexant, a buffer, a stabiliser, and a reducing agent and leads to Cu deposition on appropriate substrates immersed in the mixture. Insulating substrates (ceramic or polymer) must be activated before metal deposition can take place. A thin platinum film of only a few nanometers in thickness can act as the conductive path between the oxidation and reduction reactions, thereby enabling the formation of Cu particles or clusters in the electroless bath. Once these small metal particles or clusters are formed, the reaction becomes self-catalysed, and metal deposition will continue to cover the surface and increase the overall film thickness. The

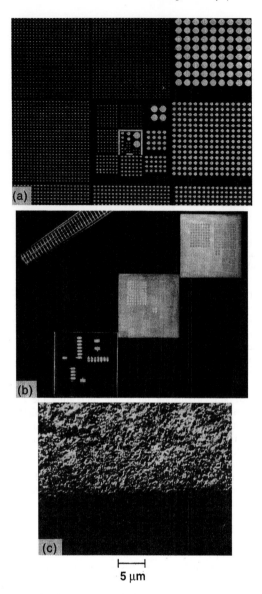

Fig. 5. Photographs and SEM micrograph showing selectivity of copper plating on AlN, Al$_2$O$_3$ and PMMA substrates after 4 min immersion in an electroless copper bath (172 nm, 50 mW/cm^2, 1 mbar, 4 min). (a) Copper patterns on Al$_2$O$_3$. (b) Copper patterns on Al$_2$O$_3$ (top), AlN (middle) and PMMA (bottom). (c) Edge quality of the copper structure.

electrical resistivity of the electroless copper was measured by using a four-point probe and found to be approximately 2.0 μΩ cm (Cu bulk resistivity: 1.67 μΩ cm).

Fig. 5 shows copper patterns produced on AlN, Al$_2$O$_3$ and PMMA substrates after 4 min immersion in the copper bath at 25°C. The largest radius on the top right hand side in Fig. 5a is 100 μm while the width of the copper lines comprising the incomplete square in the middle is 50 μm. The smallest features, just visible between most of the spots in Fig. 5a are dots with a diameter of 0.5 μm. With this process, temperature sensitive materials such as plastics, paper, polymer, and cardboard can be readily metal-coated at room temperature. Fig. 5b shows the different copper patterns on PMMA, AlN, Al$_2$O$_3$ substrates. The edge quality of these features is shown in Fig. 5c with the copper coating on the top. Clearly, this excimer lamp based approach using a quartz contact mask can produce good quality patterns. All the copper films formed on both Al$_2$O$_3$ and AlN substrates were found to have good adhesion and passed the scotch tape test.

4. Conclusions

Excimer lamp-induced activation of non-catalytic substrates using platinum acetylacetonate, for a subsequent electroless copper plating process, has been demonstrated. The decomposition rate of the platinum acetylacetonate was found to be strongly dependent upon the pressure, UV intensity and exposure time. A high decomposition rate can be obtained at low pressures and high photon fluxes. Selectively patterned copper structures with dimensions approaching the submicron level have been produced using a quartz contact mask on different substrates including temperature sensitive materials. Since larger lamps can in principle be obtained by scaling up the geometry, the large area, low temperature, and low cost VUV-induced metal deposition processes demonstrated in this work can offer interesting opportunities for copper deposition and patterning applications in ultra-large scale integrated (ULSI) devices.

Acknowledgements

This work was partially supported by EPSRC (Grant No. GR./J47750) and Brite/EuRam (PACE project No. 8073).

References

[1] Lee W, Reeves. J Vac Sci Technol A 1991;9:653.

[2] Shrivastva PB, Harteveld C, Boose CA, Kolster BH. Appl Surf Sci 1991;51:165.

[3] Lee WW, Locke PS. Thin Solid Films 1995;262:39.

[4] Hu C-K, Luther B, Kaufman FB, Hummel J, Uzoh C, Pearson DJ. Thin Solid Films 1995;262:84.

[5] Godbey DJ, Buckley LJ, Purdy AP, Snow AW. Thin Solid Films 1997;308–309:470.

[6] Esrom H, Wahl G. Chemtronics 1989;4:216.

[7] Esrom H, Zhang J-Y, Kogelschatz U, Pedraza AJ. Appl Surf Sci 1995;86:202.

[8] Esrom H, Demny J, Kogelschatz U. Chemtronics 1989;4:202.

[9] Zhang J-Y. Thesis, University of Karlsruhe, Germany, 1993.

[10] Zhang J-Y, Esrom H, Boyd IW. Appl Surf Sci 1996;96–98:399.

[11] Zhang J-Y, King SL, Boyd IW, Fang Q. Appl Surf Sci 1997;110:487.

[12] Zhang J-Y, Boyd IW. J Mater Sci Lett 1997;16:996.

[13] Lippard SJ, Morehouse SM. J Am Chem Soc 1972;94:6949.

PERGAMON

Solid-State Electronics 43 (1999) 1113–1116

SOLID-STATE ELECTRONICS

Oxidation of rf plasma: hydrogenated crystalline Si

S. Alexandrova [a,*], A. Szekeres [a], D. Dimova-Malinovska [b]

[a] *Institute of Solid State Physics, Bulgarian Academy of Sciences, Tzarigradsko Chaussee 72, 1784 Sofia, Bulgaria*
[b] *Central Laboratory of Solar Energy and New Energy Sources, Tzarigradsko Chaussee 72, 1784 Sofia, Bulgaria*

Received 17 June 1998; received in revised form 14 November 1998; accepted 16 January 1999

Abstract

The growth kinetics of thermal oxidation of H-plasma hydrogenated (100) and (111)Si surfaces is studied and is compared with that of most used so far wet chemical cleaning. Prior oxidation the silicon surface was hydrogenated by exposure to hydrogen plasma in an rf planar unit. The substrates were kept at the lower electrode without heating or at temperature of 100°C. The oxidation was performed at 800°C in dry O_2. Oxide thickness and refractive index were measured by ellipsometry. In the investigated thickness range of up to about 200 Å the growth rate was linear with time. The growth rate constant was smaller as compared to the case of Si surface with wet RCA preoxidation clean only. In spite of this, for the same time periods thicker oxides were grown on plasma treated Si substrates. Possible explanations of the experimental observations are discussed. An estimation of the film stress from variation of the refractive index over the oxide thickness is used to infer the role of the stress for the oxidation kinetics. © 1999 Elsevier Science Ltd. All rights reserved.

1. Introduction

Use of chemical passivation layers is an important issue in the control of reactive contaminants in future semiconductor manufacturing [1]. Wet chemical cleaning is known to leave a surface covered by a thin (~10 Å) layer, the so called 'native' oxide. As the dimensions shrink further into submicron and deep submicron regions, this layer comprises a significant part of the thickness of the whole active oxide. Recently, hydrogen ion bombardment has been shown to be useful as a low temperature dry cleaning technique to remove the native oxide on silicon substrates [2]. Plasma etching and cleaning in different gases has many advantages over wet chemical methods. The use of hydrogen containing plasma ambient can be of special interest for preparation of silicon surfaces for oxidation. It can be expected, however, to influence the oxidation kinetics [3].

The aim of this paper is to present results on the effect of rf plasma hydrogen passivation of (100) and (111) Si on the growth kinetics of thermal SiO_2. Silicon dioxide growth data from about 40 up to 200 Å for each orientation are carefully examined and compared with those of referent wet cleaned Si surfaces.

2. Experimental conditions

In these experiments bare 5–10 Ω cm p-type (100) and n-type (111)-oriented single-crystal Si wafers were used. All wafers were initially prepared for growing thermal oxides by a typical wet RCA pregate oxide clean [4]. Shortly after wet cleaning, some of the wafers received additionally a dry cleaning procedure by exposure to hydrogen plasma in a chemical rf planar unit. Further 'hydrogenated' and 'dry cleaned' are

---

* Corresponding author. Tel.: +359-2-778-448; fax: +359-2-754-016.
E-mail address: salex@issp.bas.bg (S. Alexandrova)

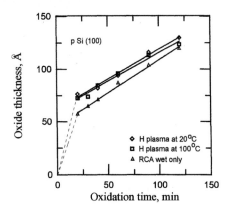

Fig. 1. Oxidation rate for p(100) Si with preoxidation exposure to H-plasma and with RCA clean only.

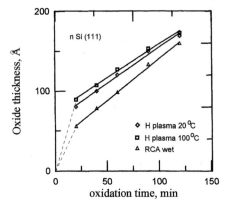

Fig. 2. Oxidation rate for n(111) Si with preoxidation exposure to H-plasma and with RCA clean only.

used interchangeably to refer the substrates with this treatment. The rf generator was capacitively coupled to the reactive chamber. The gas pressure was 1 Torr, the input power at 13.56 MHz was 15 W. The substrates were kept at the lower grounded electrode without heating or at temperature of 100°C.

Thermal oxides were grown by exposing dry and wet cleaned wafers simultaneously to dry O_2 in a conventional horizontal atmospheric furnace. During the oxidation cycle the samples were brought to the oxidation temperature of 800°C in a nitrogen ambient and then the oxygen flow was started. After oxidation the oxygen flow was discontinued and the samples were pulled to the end of the furnace and cooled there in a nitrogen flow. Oxidation times were varied in order to obtain several film thicknesses ranging between 40 and 200 Å.

Wafer thicknesses were then evaluated using a high precision Rudolf Research ellipsometer at $\lambda = 632.8$ nm. The system was approximated with a simple single-layer Si–SiO$_2$ structure although the optical constants of the SiO$_2$ layer could vary with the thickness. Standard optical constants were taken for the Si substrate. The refractive index of thin oxides since long has been known to be higher than that of the bulk [5]. Nevertheless, in many studies the refractive index has been fixed at some value, most often at 1.462 [6] or 1.465 [7], since it has been estimated that this introduces an error of ~10% or less in the thickness range of 5–40 nm [8]. In order to achieve higher accuracy which would give us a better confidence in discussing the oxidation mechanism, we preferred to determine the refractive index as well. The oxide thickness was determined with an accuracy of ±2 Å. The refractive index was correlated with the stress in the oxide film through the first-order compressibility relationship [9] $n = n_o + (\delta n / \delta \sigma) \sigma$, where $(\delta n / \delta \sigma)$ is the compressive

coefficient $(9 \times 10^{12}$ m^2 N$^{-1})$ and n_o is the refractive index of a stress free oxide. It has been shown in our previous studies that this approximation works with satisfactory accuracy for oxides with a thickness below 25 nm [10].

3. Results and discussion

The growth kinetics data for both (100) and (111) orientations are shown in Figs. 1 and 2. In all cases the oxidation is linear in time, i.e. the data can be fitted accurately to one straight line. Oxide thickness versus time plot is described by the linear equation

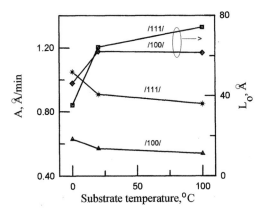

Fig. 3. Oxidation rate constant A (left y-axis) and zero time thickness L_o (right y-axis) as a function of the substrate temperature during H-plasma exposure. For reference purposes at zero temperature the data for substrates without plasma clean are given.

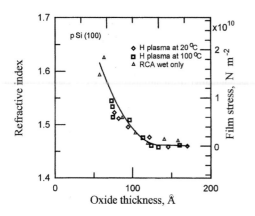

Fig. 4. Refractive index (left y-axis) and film stress (right y-axis) vs. oxide thickness for oxides grown on p(100) Si.

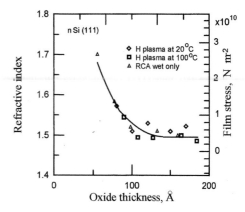

Fig. 5. Refractive index (left y-axis) and film stress (right y-axis) vs. oxide thickness for oxides grown on n(111) Si.

$D_{ox} = At + L_o$, where D_{ox} is the oxide thickness, A is the oxidation rate constant and L_o is the extrapolated value of D_{ox} to zero oxidation time. Data for A [Å/min] and L_o [Å] are given in Fig. 3.

The role of the plasma preoxidation treatment, as seen from the figures, can be regarded as follows. (a) The rate of oxidation is lower for plasma hydrogenated silicon as compared to the wet cleaned silicon both for (111)Si and (100)Si over the investigated time range. (b) The values for L_o are substantially higher for plasma-hydrogenated substrates. (c) The substrate temperature during H-plasma exposure has not a great effect on oxidation rate. All these issues will be addressed below.

Different approximation equations can be used to describe the fit of the oxidation kinetics law [7,11]. These cannot bring much clarity about the physical processes involved. However, according to the widely accepted Deal–Grove model, the kinetics of thermal oxidation of Si follows a linear–parabolic relationship [12]. The diffusion of molecular oxygen is the rate-limiting step during the parabolic regime applicable for thick films. A controversy still exists concerning the mechanism of thin-film formation, i.e. in the linear regime, which is of main concern here. The different models trying to explain this part of the oxidation kinetics have been reviewed for earlier papers in Ref. [13] and more recently in Ref. [1,14]. Although the large available experimental evidence collected so far has not allowed to accept a prevailing model, it can be stated that the role of the stress in the film and at the interface to the Si substrate cannot remain out of the main attention [1]. For that reason we have tried to discuss in some more detail this question. Higher oxide stress could mean a more dense oxide structure which would reduce oxygen diffusion thus leading to lower oxidation rate. Stress and density are related to the opti-

cal constants of the growing film. As already noted in the previous section, the behavior of the refractive index with increasing oxide thickness was evaluated and presented in Figs. 4 and 5 for (100) and (111) Si orientations, respectively. On the left hand vertical axis the refractive index and on the right hand vertical axis the film stress as a function of oxide thickness are shown, respectively. It can be seen that very close values are observed for the stress in both wet and dry cleaned silicon substrates for the whole range of oxide thicknesses. This means that at least the stress is not the limiting factor for the oxidation rate differences between wet and dry cleaned substrates.

The higher oxide thicknesses obtained for plasma hydrogenated silicon for same oxidation times have to be explained in a satisfactory manner. These are reflected in higher values of L_o in the linear approximation which describes the oxidation kinetics. The higher L_o obtained for plasma cleaned silicon indicate, however, higher growth rates in the first oxidation stages which lie outside of our experimental data. For quite rough estimation these initial growth rates are evaluated from the slope of the initial linear part, given by dotted lines in Figs. 1 and 2. These lines imply a simple linear approximation of the region below the first experimental points assuming zero oxide thickness at the beginning of the oxidation. This

Table 1
The initial growth rate A_i estimated from the slope of the linear approximation in the very initial stage of oxidation

A_i (Å/min)	(100)Si	(111)Si
RCA clean only	2.86	2.76
dry plasma clean	3.63	4.47

latter assumption is not far from reality since H-plasma has already been used to remove native oxide [15]. These oxidation rates characterizing the very initial growth A_i are given in Table 1. It can be seen that the values for the wet clean substrates are quite close for both (100) and (111) orientations. For the dry clean substrates they differ from each other and are substantially higher than those for the wet cleaned ones. The role of hydrogen here can be decisive. For to be able to approach the mechanism of oxidation the surface conditions should be known more thoroughly. Some suggestions, however, can be made even at this stage.

The silicon surface is not H-terminated after plasma treatment as in the case of some wet cleanings, since a strong inhibition of the oxidation reaction was found for such surfaces from STM observation [16]. The initial oxidation is most probably not carried out by H_2O or OH-groups since the A_i values although higher than those for wet cleaned surfaces are still not high enough to indicate mechanism similar to the wet oxidation. (Growth rates for wet oxidation are at least an order of magnitude higher than in dry O_2 [13]).

The presence of some kind of reaction layer as previously often suggested can play a significant role. It can be an oxygen-diffused zone near Si–SiO_2 interface [17]. This model explained the reversal of oxidation rates A_i of (100) and (111) Si for the wet cleaned Si (observed also by us, see Table 1 and Fig. 3) but cannot account for dry cleaned surfaces, where A_i remains higher for (111) Si. Formation of nonpassivating $Si_xO_yH_z$ layer is another possibility, as discussed in detail in the case of adsorption of H_2O at low pressures [18]. Other explanation is that hydrogen can play a catalytic role for dissociation of O_2 molecule to atomic oxygen which would increase the growth rate [1].

Before going further with the development of oxidation models for dry cleaned surfaces some issues should be addressed. Some are for example: (a) dry cleaning can influence the roughness of the silicon surface; (b) for the very initial stages of oxidation the sticking probability of oxygen is very important and it is most probably different for wet and dry cleaned surfaces. An answer of these questions should come from other measurements based on surface investigations of dry cleaned silicon surfaces.

4. Conclusions

The formation of thin oxide layers on plasma hydrogenated Si substrates is linear in time in the thickness range of about 40–200 Å. The smaller oxidation rate as compared to Si surfaces with RCA clean only cannot be directly related to the stress in the film. From the constant L_o in the linear approximation of the oxidation kinetics a fast initial oxide growth is inferred. It seems less probable that diffusion of H_2O or OH-groups are responsible for the fast initial growth. It should be a reactive layer formed by the preoxidation rf H-plasma exposure.

References

[1] Sofield CJ, Stoneham AM. Semicond Sci Technol 1995;10:215.
[2] Hu YZ, Conrad A, Li M, Andrews JW, Simke JP, Irene EA. Appl Phys Lett 1991;58:589.
[3] deLarios JM, Kao DB, Helms CR, Deal BE. Appl Phys Lett 1989;54:715.
[4] Kern W. RCA Rev 1970;31:871.
[5] Taft EA, Cordes I. J Electrochem Soc 1979;126:131.
[6] Martinet C, Devine RAB, Brunel M. J Appl Phys 1997;81:6996.
[7] Dutta T, Ravindra NM. Phys Stat Sol (a) 1992;134:447.
[8] Ajuria SA, Kenkare PU, Nghiem A, Mele TC. J Appl Phys 1994;76:4618.
[9] Fargeix A, Ghibaudo G. J Phys D: Appl Phys 1984;17:2331.
[10] Szekeres A, Christova K, Paneva A. Philos Mag Lett B 1992;65:961.
[11] Alexandrova S, Szekeres A, Koprinarova J. Semicond Sci Technol 1989;4:876.
[12] Deal BE, Grove AS. J Appl Phys 1965;36:3770.
[13] Mott NF, Rigo S, Rochet F, Stoneham AM. Phil Mag B 1989;60:189.
[14] Stoneham AM, Sofield CJ. In: Garfunkel E, Gusev E, Vul' A, editors. Fundamental aspects of ultrathin dielectrics on Si-based devices. Dordrecht: Kluwer Academic Publishers, 1998. p. 79.
[15] Rahm J, Beck E, Dommann A, Eisele I, Kruger D. Thin Solid Films 1994;246:158.
[16] Neuwald U, Hessel HE, Feltz A, Memmert U, Behm RJ. Appl Phys Lett 1992;60:1307.
[17] Murali V, Murarka SP. J Appl Phys 1986;60:2106.
[18] Ghidini G, Smith FW. J Electrochem Soc 1984;131:2924.

PERGAMON

Solid-State Electronics 43 (1999) 1117–1120

SOLID-STATE
ELECTRONICS

Thermal behavior of residual strain in silicon-on-insulator bonded wafer and effects on electron mobility

Tsutomu Iida [a,*], Takamasa Itoh [a], Yukio Takano [a], Adarsh Sandhu [b], Kazuyuki Shikama [b]

[a]*Department of Materials Science and Technology, Science University of Tokyo, 2641 Yamazaki, Noda, Chiba 278-8510, Japan*
[b]*Tokai University, 1117 Kitaname, Hiratsuka, Kanagawa 259-1200, Japan*

Received 9 June 1998; received in revised form 17 October 1998; accepted 17 January 1999

Abstract

Residual lattice strain in a bonded SOI wafer and its influence on the electron mobility were investigated as a function of heat-treatment temperature ranging from 900 to 1050°C and duration from 6 to 30 h. The change in residual strain was measured by using the X-ray diffraction method. In as-received SOI wafers, tensile strain (tensile stress along the direction parallel to the surface) of the order of 10^{-5}–10^{-4} was observed. For the specimens annealed at above 950°C, the strain varied abruptly from a tensile to a compressive one at an annealing time of 12–15 h, then approached a certain value with increasing annealing time. On the other hand, the remaining strain of the specimen annealed at 900°C exhibited the tensile region only, and did not achieve to the compressive one, even for long treatment. Significant variation in the electron Hall mobility was observed for both 1.5 and 5 μm thick SOI layers as the residual strain changes. Although samples with no strain showed identical mobility to the value in a bulk Si, the presence of the strain may anomalously affect the electron transportation. © 1999 Elsevier Science Ltd. All rights reserved.

1. Introduction

Silicon-on-insulator (SOI) is a one of the key technologies for advanced microelectronic devices [1–3]. In order to realize the reliability and performance on SOI LSIs, structural and electrical characterizations are required on the SOI-wafer in detail. In the SOI layer, residual lattice strain is existing mainly due to the differential thermal expansion coefficient between Si (2.6×10^{-6}/K) and SiO_2 (0.5×10^{-6}/K). This may give rise to instability and degradation of SOI devices. In this article, we report the behavior of lattice strain remaining in the bonded SOI layer and its influence on the electron mobility as a function of heat-treatment temperature (900–1050°C) and duration (6–30 h), and discuss formation mechanisms of the strain.

2. Experimental

Bonded n-type SOI layers 1.5 to 5 μm thick were measured in this work as listed in Table 1. These two wafers were chemically treated with aqua regia and hydrofluoric acid, then annealed between 900 and 1050°C for 6–30 h in N_2 gas ambient. Change in residual strain was measured by using the X-ray diffraction (XRD) method. Cu $K_{\alpha1}$ radiation and (400) reflection were used in the XRD experiment. Electron

---

* Corresponding author. Tel.: +81-471-24-1501; fax: +81-427-23-9362.
E-mail address: tsutomu@rs.noda.sut.ac.jp (T. Iida)

Table 1
Parameters for the SOI wafers studied in this work

	SOI-1.5	SOI-5
SOI thickness (μm)	1.5 ± 0.5	5.0 ± 0.5
SiO$_2$ thickness (μm)	1.0	1.0
Substrate thickness (μm)	550	550
SOI resistivity (Ω cm)	8–12	8–12
Substrate resistivity (Ω cm)	0.01	10
Dopant	antimony	antimony
SOI/substrate orientation	4° off from (100)	4° off from (100)

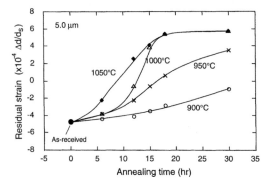

Fig. 2. Residual lattice strain in SOI-1.5 annealed from 900 to 1050°C as a function of annealing time. The positive and the negative signs of $\Delta d/d_S$ refer respectively to the compressive strain and the tensile strain.

mobility was measured using a standard van der Pauw technique at room temperature (RT).

Although the lattice planes between the substrate and the SOI layer are largely tilted to each other, we extracted the influences of the tilt successfully and measured the variation of lattice strain in the following precise manner. Let us suppose that the difference in the diffraction angles between the SOI layer and the substrate are ΔR^0 and ΔR^{180} at angles of 0 and 180° around the direction normal to the substrate surface, respectively. We can define $\Delta R^0 = \Delta\theta - \alpha$ and $\Delta R^{180} = \Delta\theta + \alpha$, where α is an tilt angle between the diffraction planes of the SOI layer and the substrate in the X-ray incident plane, $\Delta\theta$ the variation in Bragg angle due to the different lattice spacing between the SOI layer and the substrate. Since $\Delta\theta = -[(d_{SOI} - d_S)/d_S]\tan\theta$ using the interplanar spacing of the SOI layer, d_{SOI}, and the substrate, d_S, and the substrate Bragg angle, θ, the lattice strain is given by

$$\frac{\Delta d}{d_S} = \frac{d_{SOI} - d_S}{d_S} = -\frac{\Delta R^{180} + \Delta R^0}{2}\frac{1}{\tan\theta}. \tag{1}$$

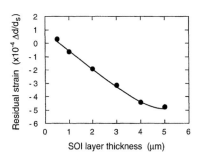

Fig. 1. Variation of the residual strain in the SOI layers as a function of layer thickness. Specimens were obtained originally from SOI-5 by etching only the SOI layer chemically.

3. Results and discussion

Variation of the residual strain as a function of SOI layer thickness is shown in Fig. 1. Specimens plotted in this figure are obtained by etching chemically from SOI-5, and not thermally treated. Accumulated strain in the SOI tends to be reduced as the layer thickness becomes thinner. It is seen that residual strain for thicker specimens tends to saturate to a certain value with increasing SOI layer thickness.

Fig. 2 shows the variation in residual lattice strain ($\Delta d/d_S$) for SOI-5 at different annealing temperatures from 900 to 1050°C as a function of annealing time. The positive and the negative signs of $\Delta d/d_S$ refer respectively to the compressive strain and the tensile strain, because the observation of the change in lattice parameter by XRD discerns the interplanar spacing normal to the substrate surface. In the as-received wafer, tensile lattice strain (lateral tensile stress applied along the direction parallel to the surface) of the order of 10^{-4} is remaining. The initial strain in the SOI layer is produced mainly by the differential thermal expansion coefficient between Si (2.6×10^{-6}/K) and SiO$_2$ (0.5×10^{-6}/K). The residual strain reduced as annealing time increases with respect to all annealing temperatures. For the samples annealed at above 950°C, the strain in the SOI layer varied from negative to positive strain at annealing times of 12 to 15 h. This demonstrates that tensile lattice strain is existing in the SOI layer for annealing times less than 15 h, then, it changes to the compressive one with increasing the annealing time. Total variation in the remaining strain for 900°C sample was rather small, showing no achievement to compressive strain even at long time treatment. For more than ~18 h, 1000 and 1050°C samples showed almost the same strain and reached a

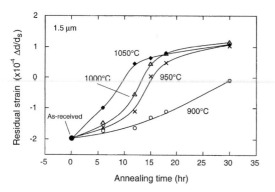

Fig. 3. Residual lattice strain in SOI-5 annealed from 900 to 1050°C as a function of annealing time.

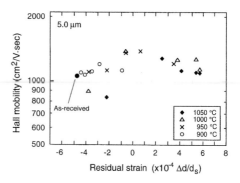

Fig. 4. Electron mobility vs residual lattice strain in SOI-5 estimated using the van der Pauw technique at room temperature.

certain value. In the case of 950°C, residual strain did not reach to the saturated value of 1000 and 1050°C even at long annealing times.

Residual lattice strain in annealed SOI-1.5 is plotted in Fig. 3 as a function of annealing time. Initial residual strain in the as-received wafer was one third of SOI-5. The total amount of change in residual strain observed for SOI-1.5 was rather smaller than that obtained for SOI-5. The substrate used for SOI-1.5 was doped with antimony (Sb) with a concentration of $\sim 5 \times 10^{18}$ cm^{-3}. This doping level brings about a lattice strain of $\sim 2 \times 10^{-5}$ [4] in the substrate. Therefore, the strain in the substrate was taken into consideration for the residual strain estimation on SOI-1.5, because the XRD measurement performed in the present work refers to the substrate lattice parameter for the $\Delta d/d_S$ evaluation. Except for the sample annealed at 900°C, it is seen that the residual strain tends to approach a certain value as annealing time becomes long. The behavior of lattice strain observed may be closely associated with the thermal change in SiO$_2$ viscoelasticity that could have a temperature dependence.

The SOI wafer is fabricated at 1100°C in O$_2$ ambient for bonding process typically, then it is cooled down to RT [5]. Assuming a good adhesion of Si and SiO$_2$ interfaces, the formation of stress is interrelated with their expansion and/or contraction as the ambient temperature changes, because of the different thermal expansion coefficients between Si and SiO$_2$. At bonding-process temperature, stresses applied to a SOI wafer (a portion of SOI layer is as thick as the substrate before thinning), SiO$_2$ and the substrate may be totally relaxed. The cooling process and subsequent mechanical thinning process for the SOI-layer formation alter the stress balance in the SOI layer, SiO$_2$ and the substrate, resulting in a change in the stress component. As a result, the as-received wafer bends

with a certain curvature, and accumulate a large amount of tensile stress in the SOI layer.

There could be two types of influences on the annealing temperature. At elevated temperatures, (i) when the annealing temperature is sufficiently high for SiO$_2$ softening, the stresses in the three layers of the SOI, SiO$_2$ and the substrate might be totally relaxed, (ii) low annealing temperature could not make the stresses relax sufficiently. Actually, it seems difficult to classify the stress conditions during annealing into the forgoing two types in an orderly manner in terms of annealing temperature and duration. For cooling down to RT, the stress due to differential thermal expansion is introduced in each layer again, then the stress balance for the layers is newly formed. It is noted that the stress applied in the SOI layer at this time is based on the new stress balance with the *thinned* SOI layer. The observed variation in the residual strain from tensile to compressive with increasing the annealing temperature could be thought as follows: the stress in the SOI layer is considered to be nearly zero due to case (i), after which, a stress introduced in the SOI may indicate a similar component to that for the shrinked SiO$_2$ as temperature descend to RT since the SOI layer is thin enough.

Figs. 4 and 5 show electron Hall mobility versus residual lattice strain in SOI-5 and SOI-1.5, respectively. It is seen that the electron mobilities have considerable relation to the presence of the residual strain. For SOI-5, while no strain sample showed identical mobility to the corresponding value of the bulk Si, the electron mobility significantly fell to one half of the maximum value with increasing the residual strain. The change in mobility for SOI-5 was smaller than that of SOI-1.5.

Strained Si film is known to exhibit a fractional change in resistivity due to the piezoresistance effect. It is considered that the change in resistivity could give

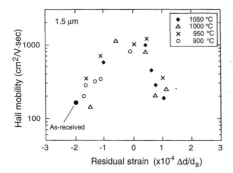

Fig. 5. Electron mobility vs residual lattice strain in SOI-1.5.

rise to a variation in the mobility. From the piezoresistance effect point of view, however, the total amount of the change in the mobility observed for the tensile or compressive regions should be much smaller than the observed values [6,7]. Therefore, it can hardly be explained that the contribution of the piezoresistance effect to the change in the mobility dependent on the residual strain is dominant. Reduced mobilities are presumably associated with (i) formation of a surface inversion layer due to the surface charge spreading effect [8], (ii) interface trapped charges, which are charged located at the Si–SiO$_2$ interface [9], (iii) fixed oxide charges, which are positive and dependent on annealing conditions [10]. These effects can also have stress dependence. Since the SOI wafer used in the present experiment is n-type conductivity, above mentioned effects might be remarkable. In order to reveal these points, extensive studies considering surface passivation with the clean process and forming a field plate to avoid surface inversion are under way.

4. Conclusion

Change in residual strain of bonded SOI wafers as a function of subsequent-heat-treatment temperature and duration was studied by means of X-ray diffraction. Residual strain in the SOI layer reduced with decreasing film thickness. The strain in the SOI layer varied from negative to positive when annealing time is beyond 15 h, indicating that the tensile lattice strain is existing for the as-received condition, then, it changes to the compressive strain with increasing annealing time. Formation of lattice strain in the SOI wafer during the fabrication process and subsequent heat treatment seems dependent on the thermal behavior of SiO$_2$ viscoelasticity. It is suggested that the change in mobility in the SOI layer is attributed not mainly to the piezoresistance effect but to other influences caused by the strain.

References

[1] Sekigawa T, Hayashi Y. Solid-State Electron 1984;27:827.
[2] Colinge JP. Electron Lett 1986;22:187.
[3] Sturm JC. Proc Mater Res Soc Symp 1989;107:295.
[4] Fukuhara A, Takano Y. Acta Cryst A 1977;33:137.
[5] Mitani K, Kanai A, Ohki K, Katayama M, Abe T. In: Schmidt MA, editor. Semiconductor Wafer Bonding: Science, Technology and Applications. Pennington: Electrochem Soc, 1994. p. 443.
[6] Pfann WG, Thurston RN. J Appl Phys 1961;32:2008.
[7] Iida T, Kanda Y, Takano Y [unpublished].
[8] Shckley M, Hooper WW, Queisser HJ, Schroen WH. Surf Sci 1964;2:277.
[9] Deal BE. IEEE Trans Electron Devices 1980;ED-27:606.
[10] White MH, Cricchi JR. IEEE Trans Electron Devices 1972;ED-19:1280.

PERGAMON

Solid-State Electronics 43 (1999) 1121–1141

SOLID-STATE ELECTRONICS

Porous silicon—mechanisms of growth and applications

V. Parkhutik

Technical University of Valencia, Cami de Vera s/n, 46071, Valencia, Spain

Received 3 July 1998; accepted 14 January 1999

Abstract

The present state-of-the-art in understanding the mechanisms of the formation of porous silicon (PS) and its physical properties is reviewed, with special emphasis on problems which were not much in the focus of existing review literature: mechanisms of the pore growth, stability of the PS properties in environment and electrical properties of PS layers. Emerging applications of porous silicon in different fields of technology are outlined. © 1999 Published by Elsevier Science Ltd. All rights reserved.

1. Introduction

During the last decade the interest of researchers to silicon, which was considered before as quite well-known material, has grown enormously. The triggering point was the paper by Dr Leigh Canham (Defense Research and Evaluation Agency, UK) who published the observation of bright red photoluminescence from the surface of electrochemically etched Si wafer [1]. Porous silicon—a substance which is produced by a treatment of Si wafers in hydrofluoric acid solutions was known since the fifties due to the works by Uhrlir [2] Turner [3], Memming and Schwandt [4]. The material was considered as suitable for electronic applications (local insulation, gettering of impurities, sacrificial layers, etc.) but never in relation with optical applications. Energy gap of silicon (1.1 eV) corresponds to the infrared region and is indirect that makes radiative recombination processes quite ineffective.

The observation of red bright photoluminescence from PS has produced a sort of sensation (although the first publication on the visible light emission from porous silicon was made in 1984 by Pickering et al.

[5]) and inspired numerous research groups worldwide to start investigations in porous silicon with the purpose of building Si-based Light-Emitting Devices (LEDs). Efforts of scientific community undertaken during the years 1991 to 1996 brought many useful results about the aspects of PS formation and its physical and chemical properties. Numerous original papers were published and further summarized in collective books, Conference proceedings and reviewing articles [6–20]. Most recent revisions of the state-of-the-art in the issue of PS are given in a comprehensive review by Cullis et al. [15] and handbook on Porous Silicon Properties edited by Canham [14], while the last international Conference on the physics and technology of porous silicon was held in March 1998 in Mallorca, Spain [16].

It appeared that the efforts of several years were not enough to start fabrication of PS-based LEDs and a certain pessimism become noticeable in considering PS as suitable optoelectronic material. Apparent problems of PS-LEDs are their low quantum efficiency, slowness and insufficient environmental stability. To resolve these problems there are still many scientific and technological issues to be addressed and studied:

1. The mechanism of PS growth still remains not very well understood. Even the latest publications dedi-

E-mail address: vitali@ter.upv.es (V. Parkhutik)

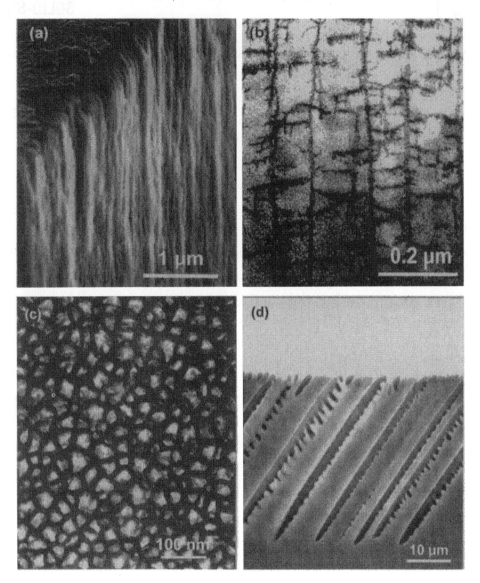

Fig. 1. Different morphologies of porous silicon films grown at p- and n-type Si. Data are taken from: (a) [24]; (b) [23]; (c) [16]; and (d) [16].

cated to this issue [14] are based on 30–40 year-old papers by Urlir [2], Turner [3], Memming and Schwandt [4], Meek [21], Unagami [22], etc.

2. Physical mechanisms of photoluminescence from porous silicon are still the subject of controversy. Although the dominating point of view is that the photoluminescence is due to radiative recombination of the carriers confined in nanoclusters of Si embedded into the walls of PS layers, there still exists the possibility that concurrent mechanisms (like luminescence of surface molecular complexes, etc.) are valid [18].

3. Mechanisms of the electroluminescence in PS are not studied in detail. The major difficulty is the absence of reliable electrical contact to the rough surface of silicon. In some cases there were indications of confusing the electroluminescence from nanostructuried clusters of silicon with other con-

current mechanisms of light emission (f.i. recombination of hot electrons in thin oxide films, etc.).

Further development of PS-LEDs and other types of devices that have emerged recently (for instance, biological and chemical sensors) would need great efforts from scientists and technologists to address the problems of physics and chemistry of porous silicon. Here we present a brief résumé of important problems for future inquiry such as: (1) Mechanism of growth and aging of PS; (2) Electrical properties; (3) Luminescence mechanisms; (4) Applications. Wherever possible, appropriate citations of preceding review literature are given for brevity and to avoid duplication.

2. Mechanisms of growth and aging

2.1. Models for the porous silicon growth

The issue of porous silicon films comprises rather different porous structures ranging from those holding micron-large pores to sponge-like layers with pores of several nanometers in diameter. Fig. 1 presents only small part of morphologies possible to observe in silicon treated in hydrofluoric acid solutions. Different structures and dimensions of pores reflect differences in preparation conditions where everything is important: type of conductivity, doping level, crystal orientation, composition of electrolyte, construction of the electrolytic cell, anodization regime, sample preconditioning and post-anodization processing, etc. In fact, samples produced by different research groups are hardly comparable even if the preparation conditions are apparently the same. No wonder a great controversy exists over the points of view of the mechanism of PS formation. Until now only a general knowledge exists about experimental parameters determining PS morphology and properties. Such issues as relative input of various reactions in the formation of PS layers, the role of anisotropy of Si wafers, the sites for the pore nucleation, the mechanisms of electric conduction at the silicon/electrolyte interface, are still to be clarified.

The mechanism of the PS growth has not attracted much attention from the scientific audience, although there is a direct relation between how PS is formed and what its properties are. It is important to ascertain that only through studying the mechanisms of PS formation can understanding of its properties exploration of the possibilities for its practical explorations be found. Here we will outline the existing mechanisms of the formation of porous structures on silicon. Earlier reviews can be found in [14] (chapter by Allongue) and [12] (chapter by Parkhutik).

The existing models can be roughly separated into three groups. The first speculates on the conditions of stability of the flat silicon surface towards the pore nucleation (initial stages of PS growth). A solid representative of these is the model by Kang and Jorne [25]. The pore nucleation at the surface of silicon is treated as a mathematical problem of the instability of a planar surface towards small perturbations. The model includes into consideration the physical processes like: (1) migration of holes and their consumption in the process of the silicon dissolution; (2) transport of reactant species (F^- ions) in the diffusion layer of the electrolyte; (3) existence of small perturbations at the semiconductor surface modeled as Fourier components $y = \xi \exp(i\omega x + \beta t)$, where ω is a spacial frequency along the perturbed surface and β represents a rate of development of the perturbation.

The analysis of the stability of the system of equations evolved shows that there are some spacial frequencies ω_{max} which correspond to maximum values of β. So, for a predetermined set of experimental conditions there should be a certain characteristic 'eigenvalue' for the preferential distance between the pores. This distance is predicted to vary as a square root of the applied voltage. Thus, the number of the active pore sites (equivalent of the density of pores) decreases with the growth of the applied potential. Other mechanisms of the pore nucleation are saturation of vacancies at the surface [26] and surface tension [27].

The second group of models deals with the stationary pore growth [28–31]. Their specifics are to consider the hole transport towards the semiconductor/electrolyte interface as a rate-determining factor. Electrochemical factors (potential-dependent side reactions, transport processes in the electrolyte, adsorption of electrolyte species etc.) are largely ignored.

The model by Beale et al. [28,32] is considered to be the first which provides quantitative estimation of the pore morphology. Its characteristic features are: (1) pinning of the Fermi level at the Si/electrolyte interface at the mid-gap due to a large number of surface states that creates a Schottky barrier between the semiconductor and electrolyte; (2) mechanisms of the current transport through the barrier are: electron tunneling for heavily doped Si and Schottky emission for lower-doped Si; (3) preferential dissolution of the pore tips is a result of a concentration of the electric field in the vicinity of hemispherical pore tips and decrease of the Schottky barrier height. Electrostatic image forces are the second reason for lowering the barrier height.

This model does not explain (neither do the majority of other models) the well-known experimental fact that the pores formed at heavily doped silicon are larger and better ordered than those grown in less-doped material. It just postulates that it is the difference in the mechanisms of conduction in heavily- and lightly-doped silicon that causes the differences in the characteristic size of the pores (6–12 and 2.5–5 nm, correspondingly).

The model by Zhang [29] allows us to calculate the distribution of the electric field in the vicinity of the pore tips and gives reasonable explanation of the localized dissolution of silicon. PS growth is postulated to occur through two competitive reactions: one through the anodic oxide growth and its dissolution and another through direct silicon dissolution in HF. The relative input of these reactions into the PS formation depends on the electric field strength: the first reaction is thought to be more probable at increased current density. The dependence of the PS morphology on anodic current density and Si doping level is explained in terms of the influence of these variables onto the potential barrier for the electron tunneling to the surface of the pore tip. A number of models employs similar assumptions of the localized electrolytic attack due to the electric field concentration at the pore tips [30] and competitive input of the reactions direct dissolution of Si or through its oxidation into the pore growth process [31].

For the p-type silicon, Lehmann and Göosele [33] developed a quantum wire model which assumes the possibility of a quantum confinement for charge carriers in the pore walls due to their very small thickness. As a result the band-gap energy increases for about 0.3 eV and penetration of electrons and holes into the pore walls becomes more difficult. The absence of holes inhibits the dissolution of the pore walls and stabilizes the porous structure.

Finally, there is also a group of synthetic models attempting to put together the stages of the pore nucleation and their stationary growth. Such are the models by Smith and Collins [34], Parkhutik and Shershulsky [35] and Makushok et al. [36].

The model by Smith and Collins [34,37] was the first allowing computer simulation of the pore morphology. The model is based on the mechanism of diffusion-limited aggregation (DLA) by Witten and Sander [38]. The branched pore growth is explained in terms of a random walk diffusion of rate-determining species (here the holes) through the depleted layer of Si to the active sites at the surface where they participate in the dissolution reaction. The process is considered similarly to the electrodeposition of metallic clusters in galvanic process (in fact, the pores are just 'negative replicas' of fibrous galvanic clusters).

Latter attempts to develop the DLA-based models [39–42] yield quite similar results, that is, computer-simulated branching of the pores but are lacking chemical specifics of PS growth. Probably, more electrochemical substantiation would be necessary before the models would be transformed from illustrative facilities into analytic tools relevant for both scientific and practical applications.

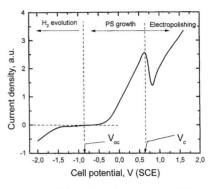

Fig. 2. Current-voltage characteristics of p-Si/HF system. V_{oc}—equilibrium potential, V_c—critical potential corresponding to the maximum of anodic current.

The model by Parkhutik and Shershulsky [35] allows both visualization and explanation of a variety of experimental facts (pore nucleation, dependence of the pore parameters on the variables of the anodization process, rearrangements of the pore structure during sharp changes of anodic potential etc.). It implies that:

1. The bottoms of the pores are covered by a virtual passive film (hereafter—VPL) which prevents a direct contact between the electrolyte and the substrate. This may be a layer of surface substance which appears as a result of a continuous dissolution of silicon and/or a double layer at the silicon surface. The concept of VPL was suggested many years ago [43] and is substantiated both by the bases of Si electrochemistry in water containing electrolyte [44] and the results of recent studies [16].

2. The pores are formed as a result of the electric field enhanced dissolution of VPL. This dissolution process may be of the electrochemical nature and involve transport of charged species (electrons, holes, ionic defects) through the electrolyte/VPL/silicon structure or proceed without the charge flow (electric field enhanced chemical dissolution—[45]). By introducing the non-linear dependence of the dissolution rate on the electric field the model goes much further than the majority of existing approaches. Fig. 2 shows well-known I-U characteristics of Si/HF system. Characteristic current peak at a potential value U_c is assumed to divide the experimental conditions into two large domains: at $U < U_c$ Si is considered to dissolve through direct transfer of Si atoms into electrolyte and it is this region where real porous silicon grows. At $U > U_c$ the dissolution is thought to proceed through the formation of intermediate oxide layer that should result in electropolishing instead of pore growth. This scheme looks oversimplified as: (I) the same

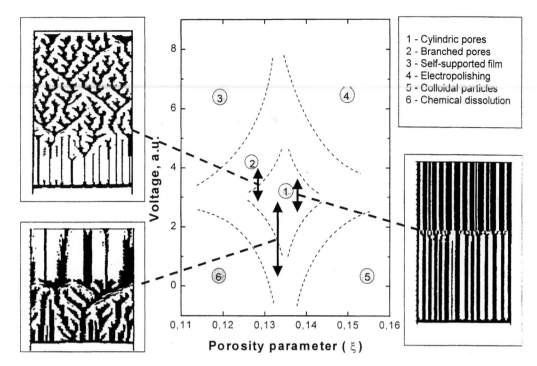

Fig. 3. Diagram 'Anodic voltage vs characteristic porosity parameter' showing possible types of morphology according to the model [12]. The inserts are computer-generated scenarios of PS morphology changes when the potential is changed within the intervals shown by arrows.

type of I-U curves is observed in the cases quite different from the present one, for example during Si anodization in acids not dissolving SiO_2 chemically [46] or in alkaline baths [47,48]. Besides, the existence of critical value of voltage U_c is not the reason but the consequence of the electrochemical and chemical processes that take place in Si/HF system.

3. The VPL growth and dissolution are determined by exponential dependences on the electric field strength in a way most generally accepted in the electrochemistry and also by factors as chemical reactivity of the material towards the electrolyte and other experimental variables [35].

The model yields a system of three-dimensional non-linear differential equations for the electric field and currents allowing both analytical solution (the steady-state pore growth) and direct computer simulation of non-stationary PS formation. Some important features of the PS morphology (like linear relationship between the size of the pores and the anodic voltage (or current); and dependence of the pore size on the electrolyte composition) are predicted by the model. Most

important, it yields a critical porosity parameter (ξ) to determine the pore morphology

$$\xi = \left[\ln\left(\frac{\alpha A}{\beta B}\right) \Big/ \ln\left(\frac{k_o}{k_e}\right) - 1 \right]^{-1} \tag{1}$$

where α and β are the coefficients establishing Faradaic relations between the current flow, on the one hand, and the VPL dissolution and growth rates on the other, k_d and k_o are exponential coefficients in the dissolution and growth terms. The establishing of the critical parameter is a real plus of this model allowing to classify and predict the possible morphologies of porous layers. The diagram in Fig. 3 shows the possible morphologies as determined by a combination of anodic voltage and the value of ξ. Marked regions (from 1 to 6) correspond to different morphology types, obtained in computer experiments but also observed in experimental conditions.

The model predicts a very attractive opportunity to modify the PS morphology by changing the anodization parameters (computer simulations of changing

Fig. 4. STM cross-sectional views of two samples of PS grown using the same Si wafer but at different current densities [49].

pore morphology are placed as inserts in Fig. 3). Just by changing the anodic voltage value one can shift the anodization process from one zone of the diagram to another and thus provoke a change in PS morphology. This possibility to change the morphology was demonstrated in direct STM measurements on PS [49]. Fig. 4 shows several cross-sectional STM pictures obtained on cleaved porous films. The samples formed on p-Si possess rather rough non-ordered porous structure (Fig. 4(a)). At the same time the pores grown at the same material at lower voltage are well-ordered cylinders lined-up with direction perpendicular to the sample surface (b).

The model by Makushok et al. [36,50] was developed to allow the analytical solution of the problem of the pore growth. The model assumes that the pore tip possesses a shape close to hemispherical. Thus the electric potential in the VPL layer may be decomposed into a series and Poisson's equation solved as a system of linear algebraic equations. A very specific feature of the model is that it does not operate with any predetermined dependence of the dissolution rate on the electric field strength. This allows to derive general conditions ensuring the growth of porous films with the predetermined morphology, as well as consider several particular cases.

The model allows to obtain a critical parameter γ which determines the morphology of the porous layers

$$\gamma(E_e) = \frac{E_e v'(E_e)}{v(E_e)} = \frac{d[\ln(v)]}{d[\ln(E_e)]} \tag{2}$$

and its critical values $\gamma_\lambda = [(r/h+1)\varphi]^{-1}$ and $\gamma_r = \varphi^{-1}$, where $v(E_e)$ is a velocity of the pore growth, $\varphi = E_{cos}/E_e$ is the coefficient determining the asymmetry of distribution of the electric field along the pore base and h is a length of the pore. Although the critical parameter γ is determined in a way different from that given by Eq.(1), it yields the same possibility to predict the pore morphology in function of the experimental conditions. The porous film would possess different types of pores or would grow as a dense non-porous layer, depending on the value of γ respectively to its critical values γ_λ and γ_r. Three possible scenarios for the pore nucleation and growth are possible.

The Si surface is stable towards the pore nucleation if $\gamma < 1$. The radii of sporadically appearing micropaths increase faster than their lengths. Here the growth of non-porous anodic films or to the electropolishing process occurs. In the case of $\gamma \approx 1.5$, the nucleation of pores takes place from the very first moments of the anodization process. Growing pores possess the bottle-like shape and their growth is hindered with time by decreasing local electric field. Finally, when the value of γ is higher than 2, the pores start to grow with their radii decreasing (needle-like

pore formation). Then at the moment when γ value approaches that of γ_r the pores change their shape to bottle-like and finally stop growing.

The model shows unambiguously that the initiation and growth of multiple pores is more probable than that of rarefied ones. Single pores would tend to form branches, as it really occurs in the case of silicon. Increase of the pore density enhances the electric field strength at the pore bottoms and its asymmetry. This allows the stable pore propagation at essentially lower values of γ. Thus, the model automatically predicts planar fronts between PS and silicon substrates. Even if there are two anodization fronts moving toward one another (as in the work by Kaushik et al. [51]), they would be planar until they merge.

The model by Makushok et al. [50] explains the larger diameters of pores and their better ordering in PS formed in heavily doped Si in terms of higher electric field strength at the pore bases and specific features of $\gamma(E)$ dependence. It predicts also the transition from the pore growth to the electropolishing with growth of the current. The model offers the possibility to determine geometrical parameters of pores, the distribution of the electric field in VPL layer and explains a variety of other experimental findings (the oscillations of the anodic current and voltage during anodization [46], the change of morphology of anodic layers from porous to barrier and different types of pores caused by the sharp changes of the anodic potential; the morphological features of the layers formed at samples with patterned surface, etc.). The model allows us to predict what type of porous structure will grow for any possible combination of substrate material and electrolyte, provided that the variables of the model may be obtained from direct experimental measurements.

Finally, it seems important to mention new mechanism of the pores growth developed recently [52]. Following the model, the pore propagation into the crystal volume is a process which involves not only electrochemical factors, but also mechanical stresses and hydrogen-related defects in Si. The role of crystal defects in the formation of PS, either generated in-situ (as the consequence of stresses acting in the PS/Si structure) or existing in Si crystal was never considered adequately as to its importance. One striking piece of evidence of the involvement of crystal defects into PS growth is the alignment of the pore nucleation sited along with the defects of the Si surface. They serve as easy paths for the pore propagation—therefore the pore growth copies the surface relief of microdefects [53].

Another important factor never considered before, is the dynamic stress accompanying pore growth. At the bottom of each pore, the dissolution reaction liberates the essential amount of hydrogen (one molecule of hydrogen per each dissolved atom of Si). Local evol-

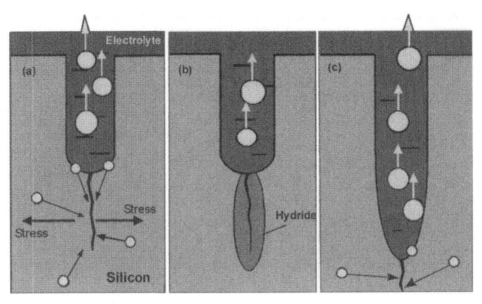

Fig. 5. Scheme of the pore growth according to the mechanical stress assisted mechanism of PS growth [52]: (a) formation of the zones of tensile stresses in underlying Si wafer and migration of hydrogen atoms into them; (b) formation of the local pockets of quasi-silicon hydride in the vicinity of the pore tips; (c) prolongation of the pore due to enhanced dissolution and cracking of the silicon hydride.

ution of Joule heat would also contribute to the formation of vapor phase [54]. Outward movement of gas bubbles and products of Si dissolution together with inward propagation of fresh electrolyte portions are to produce essential hydrodynamic pressure inside the porous silicon layer. According to the estimations made using the theory of propulsion of vapor-liquid phase in poroelastic media [55], the pressure can be as high (in the case of the pore sizes corresponding to PS) as 80–100 MPa. Thus, essential tensile stresses are produced both in porous silicon film (perpendicularly to its surface) and in Si substrates (lateral direction). Therefore, one could expect the possibility the formation of horizontal microcracks in PS (and they are seen, indeed in many PS morphologies—see e.g. Fig. 1b). Moreover, tensile stress should provoke the formation of microdefects in Si wafer right in front of each pore which are serving as easy path for further pore growth. The pressure is of dynamic character—it disappears when the process of anodization is cut off and its value depends on the applied electric field. Until now, the dominating point of view was that the stress in porous silicon is due to its contact with the environment [56]. It was assumed that the material may experience either compressive or tensile stress imposed by surface tension forces appearing as a result of the interaction of the pore wall material with absorbed gases or vapors.

According to the mechanical-stress assisted model of the porous silicon growth the development of the pore should be assumed somehow similarly to hydrogen embrittlement of the metals [57]. The stages of the process are schematically represented in Fig. 5. Hydrogen atoms, either present in the Si volume or injected from electrolyte, would flow to the vicinity of microcrack (that forms in the vicinity of the tip of the real pore due to the presence of tensile stress) to reduce their chemical potential in the tensile stress field (Fig. 5a). This results in the formation of silicon hydride at the tip of the propagating pore (b). The hydride region offers a pathway for pore propagation due to its lower mechanical stability (cleavage) and due to its enhanced chemical activity towards HF (dissolution). As a result, pore is propagating inward the Si volume following not only the direction of current lines, but also the paths of easy defect propagation in the Si crystal (usually aligned with <100> direction) and perpendicularly to it.

An important factor of the model in hydrogen is that it can either be dissolved in Si crystal or injected from the electrolyte. The role of H ions in the process of PS growth and its optical and electrical properties seems to be very important indeed. Some additional consideration of the hydrogen role will be given in section 2.2.2.

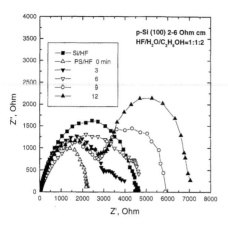

Fig. 6. Electrical impedance (complex plan view) of p-PS resting in HF solution.

2.2. Aging of porous silicon

2.2.1. Aging in forming electrolyte

Canham has suggested leaving the freshly obtained PS layer in the electrolyte to widen the pores due to chemical dissolution of Si in HF [1]. Since then this method has been widely used with the intention to dissolve Si from the pore walls. Little is known, however, about what really happens inside the pores during prolonged storage of PS in the anodizing bath.

Parkhutik et al. [16] have studied the process of PS aging in the formation electrolyte using a technique of electrical impedance which appears to be very sensitive in studying the interaction of the electrolyte with Si during its etch [58–61]. Surprisingly, the electrical impedance of PS was shown to increase with the time of its storage in HF solution, showing that some mass transfer onto the Si/electrolyte interface occurs. Fig. 6 illustrates the results for p-PS. One can see that prolonged storage of PS in HF results in appearance and development of low-frequency arc in a complex plan curves of Si/electrolyte system. Application of a short anodization current 'kills' this aging-related impedance arc but it grows again when the polarization is stopped. This feature might be related with the deposition of a passive layer at the pore bottoms, for example associated with the reabsorption of dissolved Si-complexes. These observations show the importance of accounting for the complex character of the transient chemical processes that occur at the bottoms of the pores of freshly obtained PS material.

2.2.2. Aging in environment

Due to the highly developed internal surface of PS and its reactivity towards oxygen atoms, a storage of PS films in an open ambient causes its spontaneous

oxidation. The impact of low-temperature oxidation onto photoluminescence and other properties of porous silicon is understood in general terms, although some particular features still need clarification. There are reports on both blue-shifting of PL luminescence band, no shifts or, even, red shifting [62–66] as a result of oxygen incorporation. Degradation of porous silicon luminescence in oxygen atmosphere is strongly enhanced by light illumination [66]. Other ambiences (N$_2$, H$_2$ and forming gas) did not produce essential change of PS properties [64].

Here it is very important to understand and even reconsider the role of hydrogen passivation not only in the properties of aging PS films but also in its growth (see section 2.1). It is a well-known experimental fact that the surface of freshly grown PS is passivated by the hydride layer. The main evidence for this is the observation of the IR band around 2100 cm^{-1} in spectra of PS (Fig. 7). This band is assigned to stretch vibrations in SiH$_x$ ($x = 1,2,3$) complexes at the surface of the pores [67,68]. Aging of PS is assisted by some change of the shape of the curve—small bands emerge with time at about 2250 cm^{-1} while the band at 2100 cm^{-1} is only slightly modified. The band at 2250 cm^{-1} is conventionally attributed to the vibration of Si—H bond with the surface Si atom backbonded to the oxygen one [67]. Therefore it is assumed that the oxygen is incorporated directly to the back bonds of silicon atom, while the external bonds are still saturated with H atoms.

On the other hand, it is known that the environmental aging of PS results in the appearance and enhancement of the 1100 cm^{-1} IR band due to the growth of oxide and increasing amount of Si–O–Si vibators. The growth of 1100 cm^{-1} band produces little impact onto 2100 cm^{-1} band that makes it necessary to suggest that the oxygen is incorporated to the Si backbonds without being adsorbed at the Si surface. This is a contradiction with the mechanism of the oxide growth on Si which presumes that O atoms (or ions) are to be chemisorbed at Si surface before their incorporation into the oxide [68].

The contradiction can be overcome if one takes into consideration the fact which was not used by the specialists in porous silicon, but is rather familiar to those who deal with the physics of hydrogen in crystalline silicon: namely, the same 2100–2200 cm^{-1} IR-bands which are ascribed to surface Si–H$_x$ complexes are typical for Si atoms buried into the crystal volume [69]. It was shown by application of isotopical H $\leftrightarrow$ D shifts that the complexes Vacancy-Si-H in monocrystalline defective silicon yield the structure of the IR lines very similar to what is observed in PS [70]. Then it becomes clear that the line at 2100 cm^{-1} still will survive even very harsh treatments of PS (annealing up to 500–600C) [24], chemical grafting of the surface

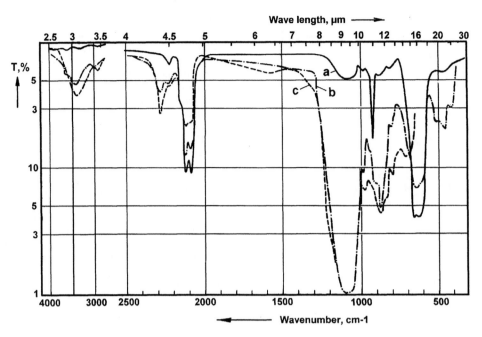

Fig. 7. (a) Infrared spectra of freshly prepared PS and after storage in air for (b) 40 and (c) 300 days [67].

[71,72], etc. These treatments leave intact Si–H bonds buried into the c–Si volume and into Si wires in the pore walls while all Si–H surface complexes at the pore walls will be substituted by silanole groups (Si–OH) or oxide molecules.

3. Electrical and optical properties of porous silicon

3.1. Electrical conductivity in porous silicon layers

Understanding of the mechanisms of the current transport through PS layers is of paramount importance to optimize the performance of PS-LEDs but still there is no clear knowledge of this important issue. There is a variety of mechanisms for the charge carrier transport through a porous silicon matrix while no dominating point of view has been elaborated so far.

3.1.1. D.c. conductivity

Ben-Chorin et al. [73] have shown that d.c. conductivity of thick PS layers is due to Poole-Frenkel mechanism of charge carrier excitation from deep traps associated with the defects at silicon nanoparticles. The same group has found strong influence of absorbed chemical species onto the d.c. conductivity [74], suggesting the possibility of intergrain transport of the carriers. Mareš et al. [75] concluded that the d.c.

conductivity of thin PS films is determined by a mechanism of tunnelling through surface barrier layer, while Kocka et al. [76] have suggested space-charge limited mechanism of the current.

The presence of surface Schottky barrier between PS and metallic contact has been frequently noted. Anderson et al. [77] measured the barrier of 0.7 eV for Al/PS dyode. Simons et al. [78] found no rectification for Al/PS structure grown at n^+-type Si, while Au–PS structures were rectifying appropriately for a Schottky barrier of about 0.74 eV. Values for the barrier height reported heretofore range from 1.2 eV [79] to 0.3–0.4 eV [80,81].

Yamana et al. [82] were first to emphasize the role which the surface plays in PS d.c. conductivity. A three-order of magnitude increase of conductivity in Au/PS/c–Si diode was observed when the relative humidity of the ambience increased from 10 to 100%. Presently, the important role of surface states in the electrical conductivity of porous silicon layers is commonly accepted. Lehmann et al. [83] attributed an increased resistivity of mesoporous silicon to narrowing pathways for the carriers in the presence of charged surface traps with activation energy about 0.5 eV. The presence of surface traps of electrons was approved by decreasing free-carrier IR absorption upon converting n^+-type silicon into PS [84]. Post-anodizing treatments, which in some way affect the con-

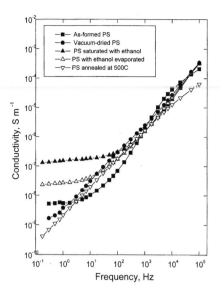

Fig. 8. A.c. conductivity of PS films subjected to different post-anodizing treatments [92].

ditions at the surface of PS, are producing dramatic effects in its electrical properties [85–89]. The problem of a stable electrical contact to the surface of PS is a most important one in measuring the environmental impacts on the conductivity. If there is no guarantee of good contact then various artifacts may be observed in conductivity data.

3.1.2. A.c. conductivity

A.c. conductivity on porous silicon in a frequency range from $1-10^5$ Hz and different temperatures (77–360 K) was reported by Ben Chorin et al. [90] and was interpreted in terms of hopping conductivity of carriers through a system of localized states with their nature remaining unspecified. Pulsford et al. [91] have performed a.c. impedance measurements on PS films (1.5 μm) contacted by various metals (Ca, Mg, Sb, Au, Ag) and shown that the PS layer formed at p^--Si forms an n-p junction with the underlying Si substrate. Parkhutik et al. [24,92] applied the impedance spectroscopy to study the post-anodizing effects in porous silicon. Fig. 8 shows the a.c. electrical conductivity (real part) vs frequency for as-formed PS sample and those subjected to different post-anodizing treatments.

The samples which are as-formed or stored in open ambient yield $\sigma(\nu)$ curves bended in a low-frequency region. By contrast, samples treated to favor the electrolyte escape from the pores (vacuum dessication at 150°C, annealing in nitrogen atmosphere, etc.) exhibit linear $\sigma(\nu)$ characteristics. It is possible to employ this difference to detect the presence of electrolyte residuals

inside the pores. For example, as-formed PS samples should clearly contain the moisture, because their conductivity is qualitatively the same as that of the samples filled-in by the electrolyte. The activation energies of 0.2–0.3 eV for a.c. conductivity in as-formed samples are too high to be attributed to hoping transport of carriers inside silicon grains. Therefore an intergrain conduction via surface traps might be responsible. Energies about 0.3–0.4 eV are similar to those reported for PS photoluminescence [93]. It seems feasible to ascertain the charge carriers in porous silicon move via hopping transport through the localized states at the surface Si according to the mechanism described in [94]. The activation energy for the electrical transport varies over a wide range (from 0.08 to 0.4 eV) and is directly influenced by the conditions at the surface of the pore walls (wet or dry, annealed or not, etc.). For this reason, both porous silicon growth and post-anodizing treatments must be strictly controlled to ensure necessary electrical properties.

The results of a.c. impedance experiments show that the routine procedures of PS finishing (washing in water and blow drying in nitrogen) are not sufficient for complete withdrawal of liquid from the pores. Provided that the surface of the sample is covered by the layer of amorphous phase [22] whose pores are narrower than those of the underlying base material, escape of the liquid will be slow and inefficient. Complete removal of the electrolyte requires deep vacuum dessication combined with sample heating. If this is not done, the electrolyte will remain inside the material, probably as a form of gel-like substance containing electrolyte components and products of Si dissolution. This substance ultimately comes to an unstable equilibrium with the porous silicon matrix. Any external factor (application of an a.c. electric field, temperature rise, contact with electrolyte, etc.) could displace it either toward additional growth of VPL or to its dissolution, thus altering the physical and chemical properties of PS layers.

3.2. Photoluminescence of porous silicon

This issue is, probably, the most extensively studied during the last years that yielded many rather different models of the luminescence from PS—see the review by Cullis et al. [15] for further details. Analysis of recent publications dedicated to studying the mechanisms of visible photoluminescence from PS shows the absence of real breakthroughs. Therefore here we will just briefly outline the present state-of-the-art in photoluminescence issues, while detailed knowledge can be obtained from preceding publications.

Following the classification introduced by Canham [14] the numerous models of the photoluminescence from PS that have appeared heretofore can be divided into four groups: (1) those based on quantum confinement (starting with the pioneering work [1]); (2) nanocrystal surface states; (3) specific defects and molecules; (4) structurally disordered phases.

Although unambiguous approval of quantum confinement model has not yet been given, it nevertheless, seems to withstand a variety of experimental facts more securely than competitive models. One of them is the observation of fine structure of luminescence edge under resonant excitation of the photoluminescence band. Calcott et al. [95] registered step-wise features in the luminescence spectra excited resonantly (excitation at wavelength corresponding to the maximum of the PL intensity) from PS at He temperatures. The steps are distant from one another for 0.57 eV that corresponds to the energy of the momentum-conserving phonons in crystalline silicon. These observations, further confirmed in many papers [96,16], show the involvement of Si crystal phonons into the luminescence mechanism and witness in favor of quantum confinement nature of luminescence.

Rosenbauer et al. [96] suggested that this fine structure of resonantly excited luminescence is due to a mechanism additional to well-known S-band photoluminescence of PS, thus arguing with the confinement-related nature of the red photoluminescence from PS.

Many experimental data, on the other hand, show the importance of the surface of Si nanoclusters in the luminescence mechanism. Surface conditions are crucial for the aging stability of the photoluminescence (see section 2.2.2). The role of surface centers within the framework of the quantum confinement theory of photoluminescence is to passivate Si dangling bonds which are efficient non-radiative recombination centers for excited hole-electron pairs in Si clusters [96]. But there is a variety of models where the surface centers are suggested as luminescent sites. For example, recent results show that selective excitation of partially oxidized PL at 2.7 eV significantly enhances the luminescent yield as compared with the material passivated by hydrogen [97]. Energy transfer processes between silicon nanoclusters and surface impurities have been registered [98]. One more circumstance that witnesses in favor of the surface-localized luminescence mechanism is a correlation of luminescent and electrical properties of PS. Both processes yield similar values of activation energy and respond synchronously to changing surface conditions [99]. Taking into consideration the very high probability of involvement of surface states into the mechanisms of the electrical conductivity (see section 3.1), the involvement of the same centers into the luminescence mechanisms may also become probable.

Further progress towards better understanding of the mechanisms of photoluminescence from porous silicon will be related with better understanding of the chemical and physical factors influencing the PS structure and properties. Particularly, efforts should be concentrated in a way to prepare material with more uniform sizes of Si clusters and better control of physical conditions at the surface of crystallites.

4. Applications of porous silicon

4.1. Optoelectronic applications

4.1.1. LEDs

The first publication on the formation of the electroluminescent devices based on porous silicon was made by Richter et al. [100] and later by Koyama and Koshida [101]. Those were Au/PS/c-Si Schottky dyodes with rather low quantum efficiency 10^{-5}–10^{-4} %. Another choice of the external contact to porous silicon was ITO [102]. Since then continuous efforts of these and many other research groups have resulted in continuous improvement of the efficiency and reliability of PS-based LEDs. Comprehensive review of the works dedicated to fabrication of LEDs can be found in the reviewing article by Cox [14].

Soon after the first attempts to build conventional Schottky type diodes, Steiner et al. [103] introduced LEDs with porous p-n contact. They have yielded increase of the efficiency up to 10^{-2}%. Loni et al. [104] have optimized this concept of LED structure and built the device with external quantum efficiency as high as 0.2% at working voltage of 2 V and continuous operation of up to 100 h in vacuum. Fabrication details could be found elsewhere [105].

The first samples of LEDs with porous p-n junctions lacked stability while working in the open environment. Linnros and Lalic [106] have shown that the environmental stability of LEDs is much improved if the pulsed mode of operation for Au/(p$^+$-n)PS/n$^+$Si/Al is used. Onset time of establishing EL was about 9 µs and decay time of about 25–30 µs, presumably due to the high impedance of the diode. Details of pulsed operation of PS-based LEDs were studied by Wang et al. [107]. Delay of electroluminescence signal from pulse-polarized diode was studied and ascribed to trapping of inhjected carriers in thin oxide layer at the Au/PS Schottky contact.

Lazarouk et al. [108,109] reported the formation of very stable Al/PS LEDs using CVD deposition of silicon onto sapphire substrates, formation of porous silicon layer, and aluminum contacting pattern (using a deposition of Al film and its local anodic oxidation). While reverse-biased, the diodes emit white light, presumably due to the excitation of surface plasmons in oxide (Al$_2$O$_3$ or SiO$_2$) near the edges of Al metallization stripes where the maximum local electric field is concentrated.

Fig. 9. S.E.M. view of a wave-guide developed on the basis of macroporous silicon layer (see [16]).

Tsybeskov et al. [110] have performed a partial oxidation of PS material to stabilise the electroluminescence yield from poly-Si/p$^+$-PS/p-PS/Si diode structure. The device has shown tunable EL yield in visible, high quantum efficiency (0.1%) at a working voltage of 2 V [111].

Although essential progress has been achieved in the fabrication of LADs based on porous silicon, further work is necessary towards increasing their stability quantum efficiency and speed.

4.1.2. Photodetectors

The formation of photoconductors and photodiodes with external quantum efficiency of 75% at 740 nm on the basis of porous silicon has been reported [112]. Photoconductors were formed of 3 μm-thick porous silicon layer on p-Si (7–10 Ωcm) by depositing aluminium film on top of porous silicon and oxidizing the structure. Its response was about 6 A/W for the incident light in 500–900 nm interval, better than UV-enhanced Si photodiode and nearly ten times exceeding

the response of similar diode prepared without thermal oxidation of porous silicon [113].

Photodiodes were formed by the electrochemical etching of p/n$^+$-Si epitaxial structure. The obtained PS layer was further subjected to RTO at 800°C for 20 s and finally Al contacts were deposited. Maximum quantum efficiency was about 75% at 740 nm. The improved passivation and lower reflectivity were suggested as responsible for high external quantum efficiency of the photodiode based on rapidly oxidized porous silicon [112].

Reduced reflectivity of incident light from rough porous silicon surface allows to use it as anti-reflecting coating for p-n photocells and enhance their spectral response and quantum efficiency [114]. To modify the spectral sensitivity of photodetectors Krueger et al. [115] suggested using multilayer PS structures working as selectively absorptive filters. The photodetector with integrated Fabry-Perot filter was selectively sensitive to predetermined wave-lengths, while without it the spectral response was rather wide in the visible range of spectrum.

Besides the applications of porous silicon as an antireflecting coating, there are some other prospects of implementation of PS in solar cell fabrication. They are considered in the reviewing article by Menna and Tsuo (see [14]). Existing research [116–118] shows that by application of porous silicon as an active layer of p-n junctions, essential improvement of solar cell performance may be achieved.

4.1.3. Photonic crystals

Photons propagating in periodic dielectric medium show the energy band structure properties similar to those of the electrons moving in periodic potential of a crystal. This type of periodic medium is called Photonic Crystals (PC) and it has attracted much research recently [119,120] due to its perspectives in optical processing of information. Existing PCs operate in microwave region, while IR and visible spectral range are still to be explored as corresponding PCs would require perfect translational order and geometrical shaping of crystals at submicron level.

Macroporous porous silicon was suggested as material suitable for fabrication of two-dimensional photonic crystals with the band gap in the infrared spectral range [121–123]. A technological scheme for optical waveguides formation uses the process of macroporous silicon fabrication developed by Lehmann [124]. Perfect arrangement of macropores is achieved by a photolithographic masking of anodized n-Si, formation of regular pattern of piramidal grooves on top of silicon (which work as nucleation sites for macroPS growth) and anodization using a back-light illumination.

A three-dimensional processing scheme was developed to form PCs on Si crystals which are 100 µm high, 2–200 µm thick and several millimeters wide [125]. It is based on passivating the pore walls by Silicon nitride, depositing Al layer on top of MPS layer and removal of unnecessary MPS by plasma etching through the windows in Al mask. Leaving some pattern nodes ungroved results in missing corresponding pores in the porous structure. (Fig. 9). These missing pores disturb a translational symmetry of photonic crystal and create localized states in the forbidden spectral region. By arranging missing pore sites into a line one can form a waveguide with rather sharp transmission band inside the forbidden gap of PC [126].

4.1.4. Waveguides

Due to porosity-dependent refractive index of the porous silicon the last may be applied to form planar optical waveguides (WG) integrated into the Si crystal. The basic requirement for WGs is that their guiding region has low optical losses and has higher refractive index than surrounding media (cladding region). There have been suggested several technological realizations of WG based on porous silicon. Bondarenko et al. [127] have developed WGs formed by local anozidation of p-Si regions and their conversion into oxide in multistage oxidation process. No specially formed cladding region was necessary as the peripherical region of oxidized porous silicon was acting as such due to its lower refraction index [128]. This scheme is popular now among those who are interested in producing PS-based WGs [16].

Loni et al. [129,130] have developed WGs by forming a PS layer of varying porosity (lower-porosity PS sandwiched between higher-porous layers and also by local thermal oxidation of the porous silicon. An advantageous feature of WGs based on oxidized porous silicon is their direct integration into the Si wafer with other optoelectronic components, fair optical losses and tolerance towards technological operations used in microelectronics. There are expectations that this type of waveguides may serve to form buried interconnections between optoelectronic components integrated into Si chips.

4.1.5. Optical logic gates

Essentially non-linear optical properties of porous silicon were explored for optical processing of signals. Matsumoto et al. [131] have proposed using free-standing PS films as active elements of all-optical integrated circuits. Absorption of incident light increased with its power. This non-linear absorption allows to achieve 50% modulation of the intensity of the secondary (reading) beam by subjecting the PS substance to irradiation by primary (writing) light beam. The switching energy is about 1 pJ thus allowing the use of this low-power operation in all-optical data processing, for example, to build logic inverters and NOR commutators. The process of induced change of absorption works with milli-second delays, indicating that rather slow localized states participate in signal transformation.

Kompan and Shabanov [132] observed a saturation of absorption of monochromatic light (633 nm line of He–Ne laser) by free-standing porous silicon films. Transparency to laser light increases with time as a result of time-dependent decrease of light absorption. This state of improved transparency is remembered by the sample for as long as one day in the dark. The memory is partially erasable by illumination of 'transparent' porous silicon by a light with shorter wavelength, thus indicating that the effect of memory is due to the processes of photocarrier redistribution between localized states rather than with illumination-assisted chemical transformations in the porous silicon.

Reflectivity of free-standing porous silicon films was

also found to experience certain memory effects: it decreases as a result of preceding illumination of the sample by laser light with 540 nm wavelength [133]. Simultaneously, interference fringes of reflected signal are blue-shifted. The memory effect here is short-lasting (around 100 ps) and is thought to be related with plasmon vibrations of short-living free photocarriers.

4.2. Microelectronic applications

4.2.1. Cold cathodes

Emission of electrons into vacuum by the surface of metals and semiconductors is considered to be a promising way of building cold cathodes of electronic components for special applications (ultrafast processing of signals, high-temperature applications or work in hazardous and radioactive environments). Active interest is attracted to the possibility of manufacturing the Si microcathodes using photolitography and chemical etching. Application of PS results are beneficial to increase the stability and performance of the Si microcathodes.

Koshida et al. [134] reported the emission of electrons from the surface of Al/n$^+$-Si/PS (40 μm) /Au structure placed into vacuum and biased in excess of 20 V (Au contact positive). Simultaneously, emission of visible light is observed from the surface. The effect is interpreted in terms of Fowler–Nordheim tunneling of electrons through the potential barrier separating PS and Au contact (presumably associated with an oxidised PS surface layer).

Wilshaw and Boswell [135] have shown that the emissive ability of cathodes formed at the Si surface by chemical etching may be essentially improved by covering it with the layer of PS. This allows to increase the maximum emission current from 1–9 μA at voltage about 1 kV between tip and anode to some 25–90 μA with voltage reduced to some 200 V. The PS cover also improves the uniformity of emission from different cathodes. The mechanism of improvement of emissive capabilities is thought to be the concentration of the electric field in the vicinity of surface microfibers which work as microcathodes at one Si tip.

4.2.2. Intercomponent isolation of integrated circuits by oxidized porous silicon

The ability of PS for fast oxidation was used to form dielectric isolation of transistors in integrated circuits. Watanabe et al. [136] developed the isolation technique called Isolation by Porous Oxidized Silicon (IPOS), which was further explored by many R&D centres [16]. The IPOS process includes basically two stages: a transformation of local regions of epitaxial Si film into porous silicon and a heat treatment of substrate in oxidizing atmosphere. Through these processes deep trenches of oxide separating pockets of

epitaxial silicon can be formed relatively easily. 6 μm deep and 6 μm wide IPOS isolation regions may be formed and withstand voltages up to 100 V at leakage currents between separated epitaxial islands of 10^{-11} A [136].

Oxidation of porous silicon, besides being a relatively fast process, presents other advantages in comparison with the process of thermal oxidation of bulk silicon crystal. The porous structure of the sample absorbs volume growth of oxide film due to Pilling-Bedworth expansion effect. Therefore, the surface of oxidized porous silicon does not contain non-planarity regions known as 'bird beaks' [137].

4.2.3. Epitaxial growth of crystalline silicon films on surface of porous silicon

Some 15 years ago, porous silicon was considered seriously as a suitable substrate material for silicon epitaxy. The idea was to obtain perfect SI crystals on top of PS and then, by opening windows in the top epi–Si, oxidize underlying PS thus achieving a complete isolation of electronic components formed in the epi–Si layer, whereas the IPOS process offered only sidewall isolation.

Unagami and Seki [138] succeeded in depositing good quality 5 μm epitaxial layer using CVD deposition process at 1170°C, beyond the temperature of PS sintering. Flawless and flat epitaxial film was obtained at a rather coarse porous silicon surface. Further research has shown that the quality of epitaxial layers depends on clearness of the PS surface, its planarity and existing strain.

Beale et al. [139] studied the deposition of MBE Si layers on top of PS formed on p-Si of both high and low resistivity. The quality of epi-Si depended essentially on the surface treatments of PS before MBE. In the case of heat cleaning at 750°C epitaxial layers showed large amount of microtwin lamelae and dislocations originating at the PS interface and protruding into epi-Si. The density of defects was reduced by preceeding sputter-cleaning of PS. Then the epitaxial layer appeared to be defectless with only occasional dislocation pairs originating from PS interface [139].

The ability of porous silicon of gettering impurities from the Si volume is considered very useful for improvement of the quality of poly-crystal epitaxial films [140].

4.3. Sensors and actuators

One of the straightforward applications of porous silicon, due to its highly developed surface and the sensitivity of its electrical, chemical and mechanical properties to external factors, is in the construction of sensors and actuators integrated into conventional Si-

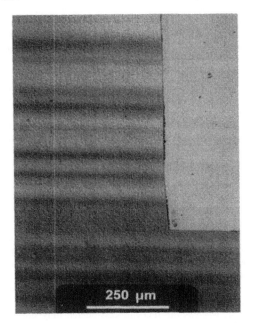

Fig. 10. Optical microphotograph of Si substrate with partially removed PS layer. Linear relief features are due to fluctuation of doping impurity over the surface of Chohralsky-grown Si and corresponding difference of PS growth rate.

based chips. Recent review of sensor applications of porous silicon has been done by Sailor [14].

4.3.1. Micromachining

Microelectromechanical systems (MEMS) are presently considered as a basis for further progress in many spheres of technology. Movable parts (cantilevers, free-standing membranes, etc.) are the basic elements of MEMS. Here PS finds growing applications as an active element of MEMS or as sacrificial layer. A brief review of technical aspects of the application of porous silicon in micromachining is made by Lang [14]. There are several reasons for PS implementation.

First, PS formation rate essentially depends on the doping level of Si wafer. Even slight variation of the doping level through Si wafer causes significant change of the PS growth rate as it is shown in Fig. 10. The highly doped p^+ and n^+ silicon is converted into porous film quite easily while low doped p and n-type silicon is much less reactive (especially in the dark). Therefore, one can convert local regions of silicon wafer into the porous layer by local doping (e.g., using ion implantation) and anodizing it. The process of PS formation stops when the anodization front contacts the interface of low-doped SI matrix.

Enhanced rate of a dissolution of highly doped silicon in HF solution was used by van Dijk and de Jonge [141] to prepare thin (0.5–20 μm) Si membranes of large surface area. Selective anodization of epitaxial Si/bulk Si structures with different doping impurity levels was further studied by Theunissen et al. [142] and Meek [21]. Since these studies, the quality of epitaxial Si films has grown considerably to ensure very good quality of thin film Si films.

Another way to produce thin Si membranes and cantilevers is by local doping of Si wafers. Very thin (0.55 μm) silicon membranes of large area (70 × 50 $μm^2$) were produced by: (1) implantation of hydrogen ions into n-Si followed by 400°C annealing in forming gas to form 3 μm thick highly doped silicon sub-surface layer; (2) implantation of nitrogen ions at 60 keV into photomask-patterned substrates to form the surface silicon nitride layer stable toward HF attack; (3) anodization of the structure in HF solution through windows left in protective silicon nitride coating to convert H-implanted layer into porous silicon which is laterally spread between undoped Si-substrate and top protective layer; (4) annealing of the structure at 1000°C in forming gas to remove the donor states associated with the implanted protons and reduce a radiative damage produced by nitrogen implantation; (5) dissolution of porous silicon regions in diluted NaOH solution [143].

The formation of PS is a process driven by the electric current. The propagation of pores is generally following electrical current lines [144]. Therefore it is possible to form zones of porous silicon buried beneath the surface of Si crystal by forcing the anodic current to pass following a required trajectory.

A second circumstance of using porous silicon in micromachining is its easy solubility in a weak (5–10%) alkaline solution while monocrystalline Si remains stable. This allows selective removal of buried PS pockets in Si volume and formation of free-standing membranes and cantilevers.

Finally, porous silicon is easily oxidized in oxygen-containing atmosphere, which is useful to produce electrical and thermal insulation of the elements of sensors and actuators from Si wafer (see [16]).

4.4. Chemical sensors

4.4.1. Sensors based on electrical conductivity effects

Both d.c. and a.c. electrical conductivity of porous silicon change dramatically as a result of the environmental impact. Moisture produces very pronounced effects in d.c. and a.c. conductivity of porous silicon (see section 3.1. for details). Absorption of other chemical species at the surface of porous silicon also produce an essential impact onto their electrical properties [145–147].

Several realizations of chemical sensors based on electrical conductivity effects in porous silicon have

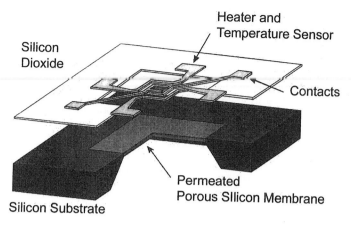

Fig. 11. Topography of gas sensor using the membrane from porous silicon [16].

been reported. Angelucci et al. [148] have fabricated sub-ppm sensor for benzene using porous silicon membrane permeated with Sn-V oxide. Its structure is presented in Fig. 11 taken from their recent work (see [16]). The sensitivity of sensor (determined as relative variation of current) was about 0.66 ppm^{-1}.

Taliercio et al. [149] have used PS membrane in a sensor of oxygen dissolved in electrolyte. PS membrane was prepared by double stage anodization of Si wafer in ethanoic HF solution. While mounted into the industrial sensor of oxygen the device demonstrated response time of about 420–600 s and permeability for oxygen about 2×10^{-10} cm^{2}s^{-1}. Optimization of the sensor performance would need a better control of the thickness of the PS membrane and its wetting ability.

4.4.2. Sensors employing photoluminescence quenching

Photoluminescence of porous silicon in contact with various vapour phases and gases may change significantly, thus opening the possibilities for optical sensing of chemical substances. Repetitive change of photoluminescence band of porous silicon from green to red and vice-versa may be observed [150] while removing the sample from an electrolyte and placing it back. Similar observations were reported by Kidder et al. [151] and Astrova et al. [152] in the case of stain etched porous silicon.

Dittrich et al. [153,154] found that the exposure of freshly formed PS films to humid atmosphere (30 Torr) at room temperature results in gradual increase of the PL yield during first two days of storage. Simultaneously the maximum spectral yield shifts towards red. Initial luminescence level may be restored by evacuating the PS sample at temperature 150°C. Treatment of PS in humid atmosphere in excess of two days results in irreversible decrease of PL intensity.

Soaking of porous silicon with chemical reagents strongly alters its photoluminescence features. Lauerhaas et al. [155] emphasized the role of dipole momentum of solvent in efficient quenching of the photoluminescence. Adsorbed molecules are thought to capture electrons or holes and reduce radiative recombination channels. Dittrich et al. [154] also studied the influence of the adsorption of oxygen, $C_2(CN)_4$ and ethyl alcohol onto the clean surface of PS layer (dried in deep vacuum at 150°C) onto its photoluminescence response. It was shown that in all cases photoluminescence is quenched and its maximum intensity is blue shifted. Adsorption of oxygen produces irreversible quenching effect due to chemical bonding of oxygen with surface silicon atoms. The effect is partially reversible in the case of $C_2(CN)_4$ and completely reversible (by vacuum drying at room temperature) for the adsorption of ethyl alcohol.

A very strong, irreversible quenching effect is produced by absorption of organoamines (f.i. $C_4H_9NH_2$) at the surface of p-type porous silicon [156]. These adsorbates are known as donors of electrons (Lewis base adsorbates) and are able to form adducts with surface active centers, thus substituting Si—H bonds.

4.5. Biological applications of porous silicon

Much effort is currently being put into integrating electronic components into biological objects. The biological incompatibility of Si material with living organisms was commonly accepted. Then Canham demonstrated that unlike the bulk Si material porous silicon becomes biologically active and resorbable [157,158]. Canham tested in-vitro the reaction of simulated body fluid with porous silicon and showed that PS stimulates a deposition of hydroappatite (main phase of bone tissues) at the surface of microporous layer. Further research has shown the prospects for the application of porous Si as material implantable into bond and that it may become an important biomaterial and reliable interface between the Si integrated circuit and biological ambient.

Bayliss et al. [159] demonstrated the suitability of nanocrystalline Si and porous silicon as growth supports for cell cultures. Their experiments have shown that the live cells are adherent and viable at the surface of porous silicon without depositing special bio-compatible coating (poly-lysine) at the surface.

Application of porous silicon in biological sensing is based on the possibility to immobilize specific biomolecules at its surface and then register changes in electrical and optical properties of PS as a result of their interaction with the complexes to be detected. Starodub [160] reported significant quenching of photoluminescence from PS as a result of antigen-antibody interaction. Photoluminescence-based immunosensor is relatively sensitive. Lin et al. [161] have developed an interferometric sensor for biological and chemical purposes. Their work is based on displacement of interference fringes in optical absorption or luminescence spectra of porous silicon selfsupporting films as a reaction for the absorption of chemical molecules or biological complexes at the pore walls. Until now the interferometric technique was applied to test DNA, antibody and biotin/avidin systems.

5. Conclusions

A résumé of the recent research and development activities in the field of porous silicon demonstrates that the material has very strong potential of practical applications, first of all in micromachining, chemical and biological sensing, while optoelectronic applications would depend to a large extent on the improvement of luminescence stability, yield and velocity.

To realize the potential advantages of porous silicon as a practically important material, further work is necessary towards better understanding of important physical and chemical factors determining its growth and properties, first of all the mechanism of the PS growth, the mechanism of the electrical conduction in PS layers and mechanisms of aging of freshly formed material.

We have confirmed recently the validity of the VPL-based model of the porous silicon growth (see section 2.1) by observing very regular and long-lasting electrochemical oscillations during Si anodization in a mixture of $0.1 \ M H_3PO_4/0.01 \ M \ HF$ acids (accepted for publication in Letters to J.Electrochem.Soc.). The oscillations are due to the repetitive formation of very thin (about 50 nm) passive nanoporous layers and

their lifting-off triggered by changing electrochemical conditions at the pore bottoms.

References

[1] Canham L. Appl Phys Lett 1990;57:1046.
[2] Uhlir A. Bell Syst Techn J 1956;35:333.
[3] Turner DR. J Electrochem Soc 1958;105:402.
[4] Memming R, Schwandt G. Surf Sci 1966;4:109.
[5] Pickering C, Beale MIJ, Robins DJ, Pearson PJ, Greef R. J Phys C 1984;17:6535.
[6] Iyer SS, Collins RT, Canham LT, editors. High Emission from Silicon. Pittsburg: Materials Research Society, 1992.
[7] Bensahel DC, Canham LT, Ossicini S, editors. Optical Properties of Low-Dimensional Silicon Structures. In: NATO ISI Series. Dordrecht: Kluwer Publishing, 1993.
[8] Fauchet FPM, Tsai CC, Canham LT, Shimizu I, Aoyagi Y, editors. Microcrystalline Semiconductors: Materials Science and Devices. Pittsburg: Materials Research Society, 1993.
[9] Tishler MA, Collins RT, Thewalt MLW, Abstreiter G, editors. Silicon-based optoelectronic Materials. Pittsburg: Materials Research Society, 1993.
[10] Vial J-C, Derrien J. Porous Silicon Science and Technology. Berlin: Springer, 1995.
[11] Collins RW, Tsai CC, Hirose M, Koch F, Brus L, editors. Microcrystalline and Nanocrystalline Semiconductors. Pittsburg: Materials Research Society, 1995.
[12] Tsu R, Feng ZC, editors. Porous silicon. Singapore: Word Science Publishers, 1995.
[13] Lockwood DJ, Fauchet PM, Koshida N, Brueck SRJ, editors. Advanced Luminescent Materials. Pennington: The Electrochemical Society, 1996.
[14] Canham L, editor. Properties of porous silicon. INSPEC, 1997.
[15] Cullis AG, Canham LT, Calcott PDJ. J Appl Phys 1997;82:909.
[16] Canham LT, Parkhutik V, editors. Porous Semiconductors—Science and Technology, 1998 To be published in J Porous Mat.
[17] Smith R, Collins T. J Appl Phys 1992;71:R1.
[18] Collins RT, Fauchet PM, Tishler MA. Physics Today 1997;1:24.
[19] Hérino R, Lang W, Münder H, editors. Porous Silicon: Material, Technology and Devices. Elsevier, 1996.
[20] Jung KH, Shih S, Kwong DL, George T, Lin T, Liu HY, Zavada J. J Electrochem Soc 1992;139:3363.
[21] Meek RL. J Electrochem Soc 1971;118:437–42.
[22] Unagami T. J Electrochem Soc 1980;127:473–6.
[23] Chuang SF, Collins SD, Smith R. Appl Phys Lett 1989;55:675.
[24] Parkhutik V, Matveeva ES, Namavar F. J Electrochem Soc 1996;143:3943.
[25] Kang Y, Jorne J. J Electrochem Soc 1993;140:2258.
[26] Corbett CW, Shereshevsky DJ, Verner IV. Physica Status Solidi 1995;147:81.
[27] Kompan ME, Kuzminov EG, Kulik V. JETP Letters 1996;64:748.

[28] Beale MIJ, Benjamin JD, Uren MJ, Chew NG, Cullis AG. J Cryst Growth 1985;73:622.

[29] Zhang XG. J Electrochem Soc 1991;140:2258.

[30] Searson PC, Macalay J, Ross FM. J Appl Phys 1992;72:253–8.

[31] Dubin V. Surf Sci 1992;274:82–92.

[32] Beale MIJ, Chew NG, Uren MJ, Cullis AG, Benjamin JD. Appl Phys Lett 1985;46:86.

[33] Lehmann V, Gösele U. Appl Phys Lett 1991;58:856.

[34] Smith RL, Chuang SF, Collins SD. J Electron Mater 1988;17:533.

[35] Parkhutik VP, Shershulsky VI. J Phys D, Appl Phys 1992;25:1258.

[36] Makushok YE, Parkhutik VP, Martinez-Duart JM, Albella JM. J Phys D, Appl Phys 1994;27:661.

[37] Smith RL, Collins SD. Phys Rev A 1989;39:5409.

[38] Witten TA, Sander LM. Phys Rev B 1983;27:5686.

[39] Erlebacher J, Sieradski K, Searson PC. J Appl Phys 1994;76:182.

[40] Sawada S, Hamada N, Ookubo N. Phys Rev 1994;49:5236.

[41] John GC, Singh VA. Phys Rev B 1995;52:11125.

[42] Weng YM, Qiu JY, Zhou YH, Zong AF. J Vac Sci and Technol 1996;B14:2505.

[43] Parkhutik V, Glinenko L, Labunov V. Surface Technology 1983;20:265.

[44] Silicon. In: Bard A, editor. Electrochemistry of Elements. Plenum, 1989. p. 234.

[45] Diggle JW. In: Oxides and Oxide Films, vol. 2. New York: Dekker, 1973. p. 281.

[46] Parkhutik V. Electrochim Acta 1991;36:1611.

[47] Palik ED, Faust JW, Gray HF, Greene RF. J Electrochem Soc 1982;129:2051.

[48] Smith RL, Kloeck B, Collins SD. J Electrochem Soc 1988;135:2001.

[49] Parkhutik V, Albella JM, Martínez Duart JM, Gómez-Rodriguez JM, Baró AM, Shershulsky VI. Appl Phys Lett 1993;62:366.

[50] Makushok Y, Parkhutik V, Martinez-Duart JM, Albella JM. In: Alwitt RS, MacDougall B, Narasubmanian R, editors. Oxides on Metals and Alloys. Pennington: Electrochem. Soc, 1992. p. 454.

[51] Kaushik VS, Datye AK, Tsao SS. Mater Lett 1991;11:109.

[52] Parkhutik V. Mater Sci Engineer B 1999;58:95.

[53] Parkhutik V, Martinez-Duart JM, Moreno D, Albella JM, Gonzalez-Velasco J. Surf Interface Analysis 1994;22:358.

[54] Sullivan JP, Wood GC. Proc Roy Soc Lond 1970;A317:511.

[55] Lee S, Salamon NJ, Sullivan RM. J Thermophys Heat Transfer 1996;10:672.

[56] Dolino G, Bellet D, Faivre C. Phys Rev B 1996;54:17919.

[57] Birnbaum HK. In: Gibala R, Hehemann RF, editors. Hydrogen Embrittlement and Stress Corrosion Cracking. American Society for Metals, 1984. p. 153.

[58] Vanmaekelberg D, Searson PC. J Electrochem Soc 1994;141:697.

[59] Ozanam F, Chazalviel J-N, Radi A, Etman M. J Electrochem Soc 1992;139:2491.

[60] Searson PC, Zhang XG. J Electrochem Soc 1990;137:2539.

[61] Koshida N, Nagasu M, Echizenya K, Kiuchi Y. J Electrochem Soc 1986;133:2283.

[62] Tsybeskov L, Fauchet PM. Appl Phys Lett 1994;64:1983.

[63] Dittrich T, Konstantinova EA, Timoshenko VY. Thin Solid Films 1995;255:238.

[64] Xu ZY, Gal M, Gross M. Appl Phys Lett 1992;60:1375.

[65] Zheng XL, Wang W, Chen HC. Appl Phys Lett 1992;60:986.

[66] Tischler MA, Collins RT, Stathis JH, Tsang JC. Appl Phys Lett 1992;60:639.

[67] Beckmann KH. Surf Sci 1965;3:314.

[68] Hasegawa H, Arimoto S, Nanjo J, Yamamoto H, Ohno H. J Electrochem Soc 1988;135:424.

[69] Bech Nielsen B, Hoffmann L, Budde M. Mater Sci and Engineer 1996;B36:259.

[70] Xie LM, Qi MW, Chen JM. J Phys, Condensed Matter 1991;3:8519.

[71] Warntjes M, Vieillard C, Ozanam F, Chazalviel JN. J Electrochem Soc 1995;142:4138.

[72] Lee E, Ha JS, Saylor M. J Am Chem Soc 1996;117:8295.

[73] Ben-Chorin M, Möller F, Koch F. Phys Rev 1994;B49:2981.

[74] Ben-Chorin M, Kux A. Appl Phys Lett 1994;64:481.

[75] Mareš JJ, Krištofik J, Pangrác J, Hospodková A. Appl Phys Lett 1993;63:180.

[76] Kocka J, Oswald J, Fejfar A, Sedlacik R. in: Proc. E-MRS Meeting at Strasbourg, 1995. (Paper I-XIV.I).

[77] Anderson RC, Muller RS, Tobias CW. J Electrochem Soc 1991;138:3406.

[78] Simmons AJ, Cox TI, Uren MJ, Calcott PDJ. Thin Solid Films 1995;255:12.

[79] Laiho R, Pavlov A. Thin Solid Films 1995;255:276.

[80] Deresmes V, Marrisael V, Stievenard D, Ortega C. Thin Solid Films 1995;255:258.

[81] Pulsford VJ, Rikken GLJA, Kessener YARR, Lous EJ, Venhuizen AHJ. J Appl Phys 1994;75:636.

[82] Yamana M, Kashiwazaki N, Kinoshita A, Nakano T, Yamamoto M, Walton CW. J Electrochem Soc 1990;137:2925.

[83] Lehmann V, Hofmann F, Möller F, Grüning U. Thin Solid Films 1995;255:20.

[84] Burrows VA, Chabal YJ, Higashi GS, Raghavachari K, Christman SB. Appl Phys Lett 1988;53:998.

[85] Möller F, Ben Chorin M, Koch F. Thin Solid Films 1995;255:16.

[85] Sabet-Dariani R, Haneman D. J Appl Phys 1994;76:1346.

[87] Motohashi A, Rike M, Kawakami M. Jap J Appl Phys 1995;34:5840.

[88] Demidovich VM, Demidovich GB, Dobrenkova EI, Kozlov SN. Sov Tech Phys Lett 1992;18:459.

[89] Schechter M, Ben Chorin M, Kux A. Anal Chem 1995;67:3727.

[90] Ben Chorin M, Moller F, Koch F. J Luminescence 1993;57:159.

[91] Pulsford NJ, Rikken GLJA, Kessener YARR, Lous EJ, Venhuizen AHJ. Appl Phys Lett 1994;75:636.
[92] Parkhutik V. Thin Solid Films 1996;276:195.
[93] Dittrich T, Timoshenko VY. J Appl Phys 1994;75:5436.
[94] Parkhutik V, Shershulsky V. J Phys D 1986;19:623.
[95] Calcott PDJ, Nash KJ, Canham LT, Kane MJ, Brumhead D. J Lumin 1993;57:257.
[96] Rosenbauer M, Finkbeiner S, Bustarret E, Weber J, Stutzmann M. Phys Rev B 1995;51:10539.
[97] Komuro S, Kato T, Morikawa T, O'Keeeffe P, Aoyagi Y. Appl Phys Lett 1996;68:949.
[98] Kenyon AJ, Trwoga PF, Federighi M, Pitt CW. J Phys, Condens Mater 1994;6:L319.
[99] Astrova EV, Belov SV, Lebedev AA, Remenyuk AD, Rud YV, Vitman RF, Kapitonova LM. Phys Stat Solidi 1994;(a)145:407.
[100] Richter A, Steiner P, Kozlowski F, Lang W. IEEE Electron Device Lett 1991;12:691.
[101] Koshida N, Koyama H. Appl Phys Lett 1992;60:347.
[102] Namavar F, Maruska HP, Kalkhoran NM. Appl Phys Lett 1992;60:2514.
[103] Steiner P, Kozlowski F, Lang W. Appl Phys Lett 1993;62:2700.
[104] Loni A, Simmons A, Calcott P, Canham LT. Electron Lett 1995;31:1288.
[105] Simmons AJ, Cox TI, Loni A, Canham LT, Uren MJ, Reeves C, Cullis AG, Calcott PDJ, Houlton MR. In: Proc. Int. Symp. on Advanced Luminescent Materials. The Electrochem. Soc. Inc, 1995.
[106] Linnros J, Lalic N. Appl Phys Lett 1995;66:3048–50.
[107] Wang J, Zhang F-L, Wang W-C, Zheng J-B, Hou X-Y, X. J Appl Phys 1994;75:1070–3.
[108] Lazarouk S, Baranov I, Maiello G, Proverbio E, de Cesare G, Ferrari A. J Electrochem Soc 1994;141:2556.
[109] Bertolotti M, Carasitti F, Fasio E, Ferrari A, la Monica S, Lazarouk S, Liakhou G, Maiello G, Proverbio E, Shirone L. Thin Solid Films 1995;255:152–4.
[110] Tsybeskov L, Duttagupta S, Hirshmann K, Fauchet P. Appl Phys Lett 1996;68:2059.
[111] Hirshmann K, Tsybeskov L, Duttagupta S, Fauchet P. Mater Res Soc Symp Proc 1997;452:705.
[112] Tsai C, Li K-H, Campbell JC, Tasch A. Appl Phys Lett 1993;62:2818.
[113] Yu LZ, Wie CR. Electron Lett 1992;28:911.
[114] Bastide S, Cuniot M, Williams P, Nam LQ, Sarti D, Levy-Clement C. in: Proceedings of 12-th EPSECE/, Amsterdam, 1994. 4A–25.
[115] Krueger M, Marso M, Berger MG, Thonissen M, Billat S, Loo R, Reetz W, Luth H, Hilbrich S, Arensfischer R, Grosse P. Thin Solid Films 1997;297:241.
[116] Menna P, DiFrancia G, la Ferrara V. Sol Energy Mater Sol Cells 1995;37:13.
[117] Schirone L, Sotgiu G, Califano FP. Thin Solid Films 1997;297:296.
[118] Duttagupta SP, Kurinee SK, Fauchet PM. Mater Res Soc Symp Proc 1997;452:625.
[119] Yablobnovitch A. Phys Rev Lett 1987;58:2059.
[120] Soukoulis CM, editor. Photonic Bandgap Materials. In: NATO ASI Series, SerE, Vol. 315. London: Kluwer Academic Publishers, 1996.
[121] Grüning U, Lehmann V. Thin Solid Films 1996;276:151.
[122] Grüning U, Lehmann V, Ottow S, Bush K. Appl Phys Lett 1996;68:747.
[123] Birner A, Grüning U, Ottow S, Schneider A, Muller F, Foll H, Gösele U. Physica Status Solidi (a) 1998;165:111.
[124] Lehmann V. J Electrochem Soc 1993;140:2836.
[125] Ottow S, Lehmann V, Föll H. J Electrochem Soc 1996;143:385.
[126] Mueller F, Birner A, Gösele U, Lehmann V, Ottow S, Foll H. J Porous Materials 1999 (in press).
[127] Bondarenko V, Dorofeev A, Kazutchiz M. Microelectr Engineer 1995;28:447.
[128] Maiello G, Lamonica S, Ferrari A, Masini G, Bondarenko V, Dorofeev A. Thin Solid Films 1997;297:311.
[129] Loni A, Canham LT, Berger MG. Thin Solid Films 1996;276:143–6.
[130] Canham LT, Cox TI, Loni A, Simons AJ. Applied Surface Science 1996;102:436–41.
[131] Matsumoto T, Hasegawa N, Tamaki T, Ueda K, Futagi T, Mimura H, Kanemitsu Y. Jpn J Appl Phys 1994;33:L35.
[132] Kompan ME, Shabanov IY. JETP Lett 1994;59:717.
[133] Bugaev A, Khakaev IA, Zubrilov AS. Optics Commun 1994;106:65.
[134] Koshida N, Ozaki T, Sheng X, Koyama H. Jpn J Appl Phys 1995;34:658.
[135] Wilshaw PR, Boswell EC. J Vac Sci Technol 1994;B12:662.
[136] Watanabe Y, Arita Y, Yokoyama T. J Electrochem Soc 1975;122:1351.
[137] Labunov VA, Bondarenko VP, Parkhutik VP, Vorozov NN. In: Proc. Electron./Electric Insulat. Confr., IEEE, Pittsburg, 1981. p. 332.
[138] Unagami T, Seki M. J Electrochem Soc 1978;125:1339.
[139] Beale MIJ, Chew NG, Cullis AG, Gasson DB, Hardeman RW, Robbins DJ, Young IM. J Vac Sci Technol 1985;B3:732.
[140] Menna P, Tsuo YS, Al-Jassimet MM. J Electrochem Soc 1996;143:589.
[141] van Dijk HJA, de Jonge J. J Electrochem Soc 1970;117:553.
[142] Theunissen MJJ, Appels JA, Verkuylen WH. J Electrochem Soc 1970;117:959.
[143] Tu X-Z. J Vac Sci Technol 1988;B6:1530.
[144] Tsao S, Myers DR, Guilinger TR, Kelly M, Datye AK. J Appl Phys 1987;62:4182.
[145] Demidovitch VM, Demidovitch GB, Dobrenkova EI, Kozlov SN. Pis'ma Zh Tekhn Fiz 1992;18:57.
[146] Stievenard D, Deresmes D. Appl Phys Lett 1995;67:1570.
[147] Motohashi A, Kawakami M, Aoyagi H, Kinoshita A, Satou A. Jpn J Appl Phys 1995;34:5840.
[148] Angelucci R, Poggi A, Cardinani GC. Thin Solid Films 1997;297:43.
[149] Taliercio T, Dilhan M, Massone E, Gyé AM, Fraisse B, Foucaran A. Thin Solid Films 1995;255:310.
[150] Li K-H, Tsai C, Sarathy J, Campbell JC. Appl Phys Lett 1993;62:3192–4.

[151] Kidder Jr JN, Williams PS, Pearsall TP, Schwartz DT, Nosho BZ. Appl Phys Lett 1992;61:2896.

[152] Astrova EV, Belov SV, Lebedev AA, Remenyuk AD, Rud YV. Thin Solid Films 1995;255:196.

[153] Dahn JR, Way BM, Fuller EW, Weydanz WJ, Tse JS, Klug DD, Van Buuren T, Tiedje T. J Appl Phys 1994;75:1946.

[154] Dittrich Th, Flietner H, Timoshenko V, Kashkarov PK. Thin Solid Films 1995;255:149.

[155] Lauerhaas JM, Credo GM, Heinrich JL, Sailor MJ. J Amer Chem Soc 1992;14:1911.

[156] Coffer JL, Liley SC, Martin RA, Files-Sesler LA. J Appl Phys 1993;74:2094.

[157] Canham LT. Adv Mater 1996;8:847.

[158] Canham LT. Thin Solid Films 1997;297:304.

[159] Bayliss SC, Harris PJ, Buckberry LD, Rousseau C. Mater Sci Lett 1997;16:737.

[160] Starodub NF, Fedorenko LL, Starodub VM, Dikij SP, Svectinikov SV. Sensors and Actuators 1996;35/36:44.

[161] Lin VS, Motesharei K, Dancil KS, Sailor MJ, Ghadiri MR. Science 1997;278:840.

PERGAMON

Solid-State Electronics 43 (1999) 1143–1146

SOLID-STATE ELECTRONICS

Two-dimensional nanowire array formation on Si substrate using self-organized nanoholes of anodically oxidized aluminum

S. Shingubara*, O. Okino, Y. Sayama, H. Sakaue, T. Takahagi

Department of Electrical Engineering, Hiroshima University, Kagamiyama 1-4-1, Higashi, Hiroshima, 739-8527, Japan

Received 18 June 1998; received in revised form 21 September 1998; accepted 2 January 1999

Abstract

A highly ordered two-dimensional array of 48 nm Cu wires was successfully fabricated on Si substrate by the usage of anodic oxidation of aluminum (Al) for the first time. Anodic oxidation was carried out for Al sputtered film on Si substrate covered by a thin thermally oxidized SiO_2 film, which was very effective to protect Si substrate from anodic oxidation. A highly ordered array of nanoholes was formed by the two steps Al anodic oxidation, and finally Cu was deposited by electroless plating in nanoholes which aspect ratio was 2.5. The present method suggests possibility of a large area two-dimensional array of quantum dots or wires on semiconductor substrate, which are considered to be a key technology for future ULSIs operated by single electron tunneling phenomena. © 1999 Elsevier Science Ltd. All rights reserved.

Keywords: Al anodic oxidation; Silicon; Nanowires; Nanoholes; Self-organization

1. Introduction

It has been discussed urgently that there are technological as well as economical limitations in lithographic technologies using optical, electron or X-ray beams for ULSI (ultra large scale integration) fabrications in the forthcoming stage of sub-100 nm scales [1,2]. For this reason, much attentions has been paid for nanostructures formation by self-organizing methods such as strain-induced quantum dots formations [3,4], nanocrystal formation on the atomic step edges [5–7] and nanoholes formation by Al anodic oxidations [8–11].

---

* Corresponding author. Tel.: +81-824-247-645; fax: +81-824-227-195.
E-mail address: shingu@ipc.hiroshima-u.ac.jp (S. Shingubara)

Among these methods, Al anodic oxidation has been shown to be capable of realizing an extremely highly ordered periodic structures of nanoholes by the usage of the two steps anodization [9,10]. The authors recently showed that metallic wires array could be formed by electroplating in nanoholes on Al plate [11]. However, extension of nanohole array formation to semiconductor single crystalline substrates such as Si and GaAs is required for a wide applications to microelectronics. The aim of the present study is to form two-dimensional (2D) array of quantum dots and wires on Si substrate by the usage of Al anodic oxidation. For this purpose, electroless plating is carried out to grow nanowires in Al anodic nanoholes on Si, and furthermore primary investigations of selective deposition of semiconductors as well as metals are discussed.

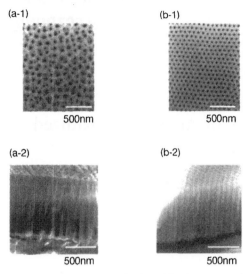

(a-1)   (b-1)

500nm   500nm

(a-2)   (b-2)

500nm   500nm

Fig. 1. SEM images of nanoholes formed by the two steps Al anodic oxidation at a condition far from self-organization (a) and at a self-organization condition (b). (a-1) and (b-1) are views from the top, and (a-2) and (b-2) are cross-sectional views. Conditions of anodic oxidations are; (a-1, a-2): oxalic acid concentration 0.15 M, 60 V, 5°C, 1st step anodization time is 25 h, 2nd step anodization time is 15 min. (b-1, b-2): oxalic acid concentration 0.5 M, 40 V, 5°C, 1st step anodization time is 32 h , 2nd step anodization time is 30 min.

2. Experimental results

For anodic oxidation of Al, oxalic acid solution of 0.1–0.6 M was used. Array of holes, which are perpendicular to the interface between aluminum and aluminum oxide, develops along with the growth of aluminum oxide. It is reported that at some conditions of temperature, voltage and anodization time, an excellent ordered two-dimensional array of nanoholes could be obtained [9,11]. Arrangement of holes are delicately dependent on the initial surface roughness, and there is a tendency to align into ordered structure when holes grow in vertical direction during long time anodic oxidation. In order to realize two-dimensional ordered array of holes, two steps anodic oxidation was proposed [9,10]. The aluminum oxide film is removed after the first step anodic oxidation with enough long time by wet chemical etching using phosphorous acid, then remained aluminum surface roughness reflects two-dimensional array of bottoms of holes. An ordered array of holes is obtained after the succeeding second anodic oxidation. Fig. 1 shows structure of nanoholes which were formed in Al plate after the two steps anodization at different voltages. An excellent 2D ordered array of nanoholes was formed when $V = 40$ V. On the

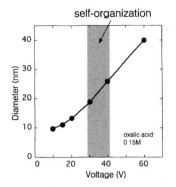

Fig. 2. Average diameter of holes as a function of the voltage of anodic oxidation. The oxalic acid concentration is 0.15 M, and the temperature is 5°C. Self-organization of nanoholes are obtained at a voltage between 30 and 40 V by the two steps anodization.

other hand, when $V = 60$ V, arrangement of nanoholes is random and cross-sectional image showed branching of holes during oxidation. Average diameter of nanoholes are 25 and 40 nm for 40 and 60 V, respectively. Thus self-organization of nanoholes is dependent on the voltage. Fig. 2 shows voltage dependence of nanohole diameter. Mean hole diameter is monotonously increased with increasing the voltage, as it has been reported formerly [8]. Self-organization was observed between 30 and 40 V at rather high oxalic acid concentration at temperature around 5°C.

We have investigated nanoholes array formation on Si substrate. Pure Al (99.999%) was deposited on Si substrate covered by a thin (30 nm) thermally oxidized SiO_2 film by DC sputtering. When there was no SiO_2 between Al and Si-substrate, anodic oxidation was not stopped at the Al/Si interface and porous Si was grown finally. It is desirable to limit anodic oxidation within Al layer in order to make electronic contact between substrate Si and quantum dots or quantum wires formed in nanoholes. SiO_2 worked a very good barrier to anodic oxidation as well as wetting layer between porous alumina and Si. At first we have tried to use gold (Au) film as an barrier for anodic oxidation, however, there was a problem of adhesion between porous alumina and Au. Then we have tried to deposit metals such as Ni and Cu by plating in the nanoholes.

Fabrication sequence of nanoholes and nanowires are shown in Fig. 3. Two steps anodic oxidation on Si is schematically shown in Fig. 3(a–d). In order to deposit metals in the holes, Al film thickness after removal of 1st step anodized Al should be smaller than a few hundreds nm. We have deposited thick Al film of 10–30 μm initially, then the first step anodization was carried out until the rest Al film thickness

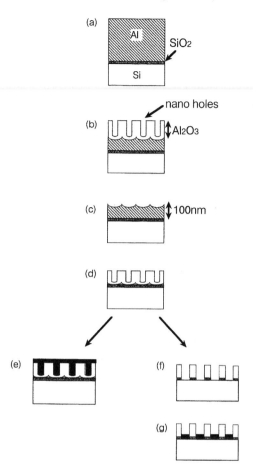

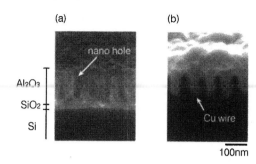

Fig. 4. Cross-sectional SEM views of nanoholes formed by the two steps anodization (a) and Cu nanowires deposited by electroless plating. Thickness of SiO_2 is 30 nm. Al anodization condition was 40 V, 0.3 M. Alumina was slightly etched away by 5 wt% phosphoric acid before electroless plating. Mean diameter of Cu wires is 48 nm, and aspect ratio is 2.5.

Fig. 3. Fabrication process sequence of nanowire and/or dots array on Si substrate by the Al anodic oxidation. (a) Pure Al film with a thickness of 10–30 μm is deposited on a heavy doped Si substrate covered by a thin silicon oxide. (b) The first step Al anodic oxidation is carried out at the self-organization condition until the rest Al film thickness is as small as 100 nm in order to lower aspect ratio of nanoholes than 5. (c) Alumina is etched away by the mixture of phosphoric acid and chromic acid. (d) The second anodic oxidation is carried out until Al is completely converted to porous alumina. (e) Deposition of metal and/or semiconductor by nonselective deposition such as electroless plating to form nanowires array. (f) Reactive ion etching (RIE) of the bottom alumina barrier layer of nanoholes, and subsequent RIE of SiO_2 to open windows for substrate Si. (g) Selective deposition of metal and/or semiconductor to nanoholes. By adequate controll of deposition thickness, nanowires as well as dots can be fabricated.

nanoholes; selective or nonselective. We have investigated both by plating method and succeeded in deposition of Cu in the nanoholes by nonselective electroless plating as schematically shown in Fig. 3(e). Fig. 4 shows cross-sectional view of nanoholes on Si substrate (Fig. 4a), and deposited Cu by electroless plating. The surface of Al-oxide was activated by $PdCl_2$ treatment prior to the electroless plating using $CuSO_4 \cdot 5H_2O$. When the aspect ratio (ratio of the diameter to the height) of the hole was 2.5, the hole was completely filled by Cu as shown in Fig. 4(b). However, it was difficult to fill Cu completely when aspect ratio was larger than 5. Two-dimensional array of Cu nanowires with 48 nm diameter thus obtained is shown in Fig. 5. An excellent 2D array was obtained as if they were formed by lithographic technology.

Electroplating is promising as a selective deposition

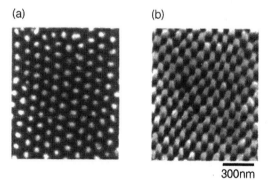

Fig. 5. SEM photographs of two-dimensional array of Cu nanowires. The mean diameter of Cu wires is 48 nm. (a) Top view. (b) Bird's eye view. Cu film with nanowires array was delaminated by scotch tape pulling from Si substrate for SEM observation.

was as large as 100 nm. After the second step anodization of Al, Al film was completely changed to Al-oxide. There are two choices of material deposition to method. We have to remove both bottom barrier layer of Alumina of nanoholes as well as SiO_2 film (Fig. 3f and g) for this case. At first we have tried to etch away the bottom Al-oxide by Ar ion beam etching, however we have failed since the top of holes were closed due to redeposition of Al-oxide. Further study for anisotropic etching of the bottom Al-oxide and SiO_2 by RIE (reactive ion etching) are in progress.

3. Concluding remarks

Two-dimensional array of Cu nanowires of 48 nm diameter is successfully fabricated by two steps anodization of Al which was deposited by sputtering on Si substrate covered by thin SiO_2. The present method is capable to fabricate ordered nanowires array of a variety of materials, and further shrinkage of wire dimensions would be necessary to application to the nanofabrication of quantum materials and devices which are operated at room temperature. The shrinkage would be possible by searching a self-organization condition at the lower anodic voltage. Selective deposition of metals as well as semiconductors in nanoholes is essential for this final goal, and further investigation to remove the bottom Al_2O_3 barrier layer by RIE using chlorine-based reactive gases is in progress.

Acknowledgements

The authors would like to express gratitude to Professor H. Masuda of Tokyo Metropolitan University for kind instructions and suggestions of two steps anodization of aluminum. This work has been supported by CREST (Core Research for Evolutional Science and Technology) of Japan Science and Technology Corporation (JST).

References

[1] Canning J. J Vac Sci Technol B 1997;15:2109.
[2] McCord MA. J Vac Sci Technol B 1997;15:2125.
[3] Jesson DE, Chen KM, Pennycook SJ. MRS Bull 1996;21-4:31.
[4] Petroff PM, Medeiros-Ribeiro G. MRS Bull 1996;21:50.
[5] Kawasaki K, Mochizuki M, Tekeshita J, Tsutsui K. Jpn J Appl Phys 1998;37:1508.
[6] Sakaue H, Katsuta Y, Konagata S, Shingubara S, Takahagi T. Jpn J Appl Phys 1996;35:1010.
[7] Uejima K, Takeshita J, Kawasaki K, Tsutsui K. Jpn J Appl Phys 1997;36:4088.
[8] Keller F, Hunter MS, Robinson DL. J Electrochem Soc 1953;100:411.
[9] Masuda H, Fukuda K. Science 1995;268:1466.
[10] Masuda H, Satoh M. Jpn J Appl Phys 1996;35:L126.
[11] Shingubara S, Okino O, Sayama Y, Sakaue H, Takahagi T. Jpn J Appl Phys 1997;12B:7791.

PERGAMON

Solid-State Electronics 43 (1999) 1147–1151

SOLID-STATE ELECTRONICS

Coulomb blockade: Poisson versus Pauli in a silicon quantum box

L. Palun [a,*], G. Lamouche [b], G. Fishman [b]

[a]*Laboratoire d'Electronique de Technologie et d'Instrumentation, Commissariat à l'Energie Atomique, 17 Rue des Martyrs, 38054 Grenoble Cedex 9, France*
[b]*Laboratoire de Spectrométrie Physique, UMR C5588, Université Joseph Fourier – Grenoble 1, CNRS, BP 87, 38402 Saint-Martin d'Hères Cedex, France*

Received 15 June 1998; received in revised form 29 November 1998; accepted 11 January 1999

Abstract

We discuss the limitations of the orthodox Coulomb-blockade theory when applied to silicon quantum dots in the nanometer range and we present a simple Poisson–Schrödinger model to evaluate the quantum contribution in these cases. This contribution can be seen as a quantum capacitance in series with the sum of capacitance around the dot. This simple model gives results similar to a more sophisticated one which includes Pauli principle, with a precision of the order of room-temperature thermal–energy kT. Finally we show that the simple model can be easily included in micro-electronic simulators and therefore can be very effective to predict new properties of future quantum devices. All the effects discussed in this paper are room-temperature effects. © 1999 Elsevier Science Ltd. All rights reserved.

1. Introduction

Devices using Coulomb-blockade effect, like metallic single electron transistors, are well-known [1]. For an industrial application, those devices must have MOSFET compatibility and must operate at room temperature. These two conditions have to be fulfilled in silicon devices. In fact, the Coulomb energy $e^2/2C$ must be five-to-ten times larger than room thermal energy kT to be effective as a blocking mechanism [2,3]. This implies devices based on quantum dots, namely nanometer dots. The Coulomb-blockade charging effect of a few nanometer silicon dot does not only result from capacitors around the dot, but also from the quantization of energy-levels. Consequently,

conductance in such a device shows a parity in oscillations. This phenomenon is described in Ref. [4].

We present here a simple model to describe one and two electrons in a silicon dot. This model combines (i) the effective-mass approximation for particles, (ii) a square potential for silicon dots and (iii) a Poisson approach for electrostatic potential. In other words the energy quantization is calculated via a Poisson–Schrödinger equation. This allows us to describe additional energy due to level quantization. Fig. 1 shows that it is sufficient to consider one electron or at most two electrons in nanometer silicon devices.

For Coulomb-blockade effect calculation in nanometer silicon dot, the level quantization is modeled as an internal capacitor in series with the sum of junction and gate capacitors around the dot.

This simple model is then compared with a more sophisticated one including Pauli principle [5], where the problem is exactly solved in a parabolic potential.

---

* Corresponding author. Fax: + 33-3-88837892.
  E-mail address: palun@sorbier.cea.fr (L. Palun)

0038-1101/99/$ - see front matter © 1999 Elsevier Science Ltd. All rights reserved.
PII: S0038-1101(99)00038-6

Nomenclature

m_{Si}, m_{SiO_2}	effective mass for electrons in Si and SiO_2 in unity of electron mass
C_Σ	sum of all junctions and gate capacitance
C_i	internal capacitance of silicon dot
a	radius of silicon dot
r_0	typical radius for the 0s wave function in a parabolic potential
ε_{Si}, ε_{SiO_2}	silicon and silicon–dioxide dielectric constant
ε^*	effective silicon dielectric constant for parabolic potential
ω	parabolic–potential well frequency

2. Silicon dots modeling

Spherical silicon quantum dots studied in this paper are undoped, crystalline and surrounded by silicon–dioxide. They are about 2–12 nanometer large. In following we are interested by the conduction-band electrons.

Calculations are made with an effective mass $m_{Si} = 0.27$ for electrons in Si, corresponding to $(2/m_t + 1/m_l)/3$, where m_t and m_l are respectively the transversal and the longitudinal effective-masses. This mass is consistent with the average mass that can be used to calculate the energy in a cubic infinite-square quantum well. The electron in SiO_2 is considered having an effective-mass of $m_{SiO_2} = 0.5$, assuring SiO_2 is a crystalline semi-conductor [6].

The dot is first modeled by a spherical finite–square–potential well. The two parameters of the well are the radius a and the depth V_0, with $V_0 = 4$ eV [7]. Even if it is a very high barrier, the infinite-well model

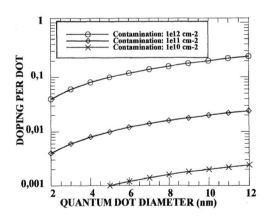

Fig. 1. This figure shows the mean dot–equilibrium–charge-state for 3 surface contamination: 10^{10}, 10^{11} and 10^{12} impurities per cm^{-2}. Dots are spherical. The distance between dots is 2 nanometers. Every surface impurity is assumed to give one charge carrier.

is not an efficient model because the fundamental level can be high too, typically 0.1–1 eV.

For one electron in the well, fundamental level E_{0s} is extracted from Schrödinger equation $H\Psi_{n,l,m} = E\Psi_{n,l,m}$, where $H = p \cdot m^{-1}p + V$, with m equal to m_{Si} or m_{SiO_2}. Since we are only interested in the fundamental level E_{0s}, the problem boils down to one-dimensional. E_{0s} is evaluated by insuring the continuity of wave function ψ_{0s} and the continuity of probability current J_{0s} at the Si–SiO_2 interface Eq. (1).

$$1 - ak_1 \cot(ak_1) = \frac{m_{Si}}{m_{SiO_2}}(1 + ak_2). \tag{1}$$

Wave vectors k_1 and k_2 are defined respectively in Si and SiO_2 (Eqs. (2a) and (2b)).

$$k_1 = \sqrt{\frac{2m_{Si}E_{0s}}{\hbar^2}}, \tag{2a}$$

$$k_2 = \sqrt{\frac{2m_{SiO_2}(V_0 - E_{02})}{\hbar^2}}. \tag{2b}$$

When a second electron is added, a new term appears in the Hamiltonian H, corresponding to the interaction between the two carriers. In the simple model first considered, the electronic cloud of one electron can be considered as a charge density $\rho(r) = q|\psi_{0s}(r)|^2$ by the other one, where q is the elementary electron charge. This charge density induces a potential modification taken into account in Poisson equation Eq. (3), where $\varepsilon_{rel} = \varepsilon_{Si}$ for $r < a$ and $\varepsilon_{rel} = \varepsilon_{SiO_2}$ for $r > a$.

$$\Delta V_P(r) = -\frac{q}{\varepsilon_0 \varepsilon_{rel}}|\psi_{0s}(r)|^2. \tag{3}$$

The consequence of this interaction is a shift ΔE_{0s} on fundamental level. Fig. 2 illustrates those considerations.

With the Poisson–Schrödinger approach, the Pauli principle is not applied. The energy calculation taking

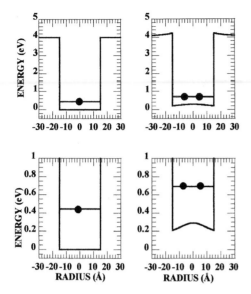

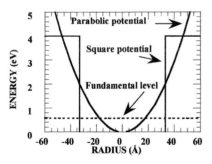

Fig. 2. Potential well and energies modifications when an electron 'sees' another electron. Electron wave function leads to a charge density into Poisson equation. The lower figure is a larger view of the upper one to print out the energy shift between the well with one or two electrons.

into account the Pauli principle for two interacting electrons has been carried out in three-dimensional parabolic potential $m_{Si}\omega^2 r^2/2$ [5]. Scaling such a potential with the square one as shown in Fig. 3, it is possible to look at the error made without taking into account the Pauli principle. The two parameters for scaling are the well frequency ω and the parabolic-well silicon–dielectric–constant.

The parameter ω is chosen to lead to the same fundamental levels E_{0s} as for the square-well potential.

Fig. 3. Parabolic potential compared to square potential. The square potential figures a silicon quantum-dot surrounded with silicon dioxide. The parabolic potential figures the electrostatic potential which is used on the present paper taking into account the Pauli principle.

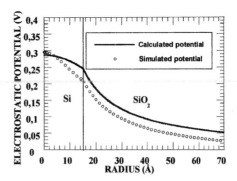

Fig. 4. Electrostatic potential induced by an electron in a silicon quantum-dot surrounded with silicon dioxide. (○): the charge density is proportional to $|\psi_{0s}(r)|^2$. (−): the charge density is uniform inside the dot and zero outside.

The relation is well known: $E_{0s} = {}^3/_2 h\omega$. For a tractable comparison, the value of the electrostatic potential at the center of the dot due to charge density induced by an electron must also be equivalent to that of the square-potential well. It is possible to introduce an effective silicon-dielectric-constant ε^* which takes into account the effect of the confinement of electron wave function in the real quantum-dot on the approximate interaction potential. This effective value can be evaluated analytically. Approximating the charge distribution by uniform charge densities within spheres of given radii in the square well and in the parabolic well, the Poisson equation allows one to calculate the electrostatic potential modifications. For the square potential, the charge density is taken as uniform in the dot and zero outside. Fig. 4 shows the electrostatic potential induced by this model compared with one obtained from the true one-electron wave function. For the parabolic potential, the charge density is assumed uniform within a $2r_0$-radius-sphere, where the characteristic length $r_0 = (h/m_{Si}\omega)^{1/2}$ is related to the 0s wave function Eq. (4).

$$\psi_{0s}(r) \propto \exp\left(-\frac{1}{2}\left(\frac{r}{r_0}\right)^2\right). \qquad (4)$$

Insuring that the two electrostatic potentials (approximate square-well and parabolic potentials) take the same value at the center of the dot, the effective dielectric constant ε^* can be evaluated Eq. (5).

$$\varepsilon* = \frac{3}{2}\varepsilon_{Si}\frac{a}{r_0}\left(1 + 2\frac{\varepsilon_{Si}}{\varepsilon_{SiO_2}}\right)^{-1}. \qquad (5)$$

The parabolic potential gives a physical description less faithful than the square one, but it is much easier to handle.

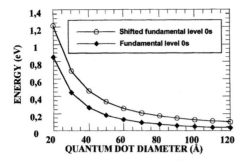

Fig. 5. Fundamental 0s levels for one and two electrons in a silicon quantum-dot modeled by a square well potential. The 0s level shift for two electrons is calculated by a consistent Poisson–Schrödinger simulation. The well depth is 4 eV. The effective-mass for an electron is 0.27 in silicon and 0.5 in silicon dioxide.

3. Quantum capacitance modeling

In a metallic single-electron device, Coulomb blockade energy E_{CB} is written $e^2/2C_\Sigma$. In more-than-twenty-nanometer semiconductor devices, particularly when the dot is defined by a strong electrostatic confinement at the interface of two semiconductor layers, an internal capacitance C_i must be taken in account in the description of the Coulomb blockade. C_i is defined via the energy which is necessary to bring an electron from infinite to the electron gas of the dot. C_i is put in series with C_Σ. For a 'non-quantum' spherical dot of radius a, $C_i = 4\pi\varepsilon_{Si}\varepsilon_0 a/3$.

In a silicon quantum dot, C_i has no meaning with an electron–gas model because of the very little number of charge carriers in the conduction band. C_i de-

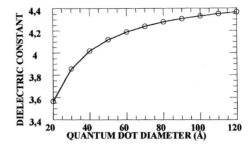

Fig. 7. Effective dielectric constant ε^* used in the parabolic potential Poisson–Schrödinger simulation. $\varepsilon^* = 3a\varepsilon_{Si}/2r_0(1 + 2\varepsilon_{Si}/\varepsilon_{SiO_2})$, in which a is the quantum dot radius and $r_0 = v(h/m_{Si}^*\omega)$ characterizes the extension of the wave function in the parabolic potential. The dielectric constants are respectively 11.9 and 3.9 for silicon and silicon dioxide.

rives from the level shift ΔE_{0s}. Its value is $C_i = e^2/2\Delta E_{0s}$.

E_{CB} is written $e^2/2C_{total}$, with $C_{total}^{-1} = C_\Sigma^{-1} + C_i^{-1}$: when quantum capacitance decreases, Coulomb blockade plateaus increases.

4. Results

Fig. 5 shows the fundamental 0s levels versus dot diameter. The zero reference for energies is the bottom of the conduction band. If we consider that Fermi level E_F in the dot is fixed by silicon electrodes around the dot and introducing the 0s level for one electron in Fermi–Dirac statistic, it confirms that every dot contains at most one electron in the conduction band at equilibrium and without source–drain applied voltage.

From the 0s level shift a quantum capacitance is

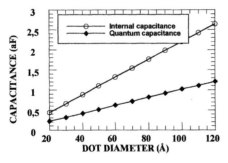

Fig. 6. Quantum capacitance of a silicon quantum-dot (♦) compared to classical internal capacitance of a spherical electronic gas (○). The quantum capacitance is calculated from the level shift, with $\Delta E_{0s} = e^2/2C_{quantum}$. The internal capacitance is defined by $C_{internal} = 4\pi\varepsilon_{Si}\varepsilon_0 a/3$, in which a is the silicon dot radius and $\varepsilon_{Si} = 11.9$. As for classical approach, the law for quantum capacitance versus dot diameter is linear.

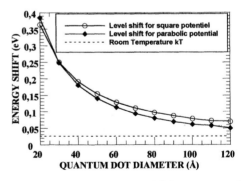

Fig. 8. Comparison of the 0s level shift in square potential (○) and parabolic potential (♦). The two shifts differ by less than room temperature thermal energy kT, which is a sufficient precision for micro-electronic application allowing for other uncertainty, as dot geometry.

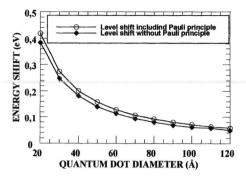

Fig. 9. Comparison of the 0s level shift taking into account Pauli principle (○) and consistent Poisson–Schrödinger calculation (◆) for a parabolic potential. The difference between the two shifts is less than room-temperature thermal-energy kT.

extracted (Fig. 6). As for classical internal capacitance, the variation of the quantum capacitance versus dot diameter is linear. If we consider that the quantum effects are detectable for a level shift higher than five times the room temperature thermal energy kT, or higher than 0.6 aF in term of capacitance, devices must integrate under-six-nanometer-size silicon dots. This figure also shows the limitation of classical calculations: an error by a factor 2 error on the capacitance value induces the same error on Coulomb plateaus.

This model does not take into account Pauli principle. Modeling the dot by a parabolic-potential well, according its frequency ω and introducing an effective dielectric constant ε_{Si}^* (Fig. 7) to have the same E_{0s} and electrostatic potential at the center of the dot (Fig. 8), it is possible to avoid the error made with Poisson–Schrödinger resolution (Fig. 9). This error is less than 9% for a 2-nanometer-diameter dot and increases to 15% for a 12-nanometer-diameter dot. Indeed, the repulsion effect becomes more prominent with increasing of size dot.

5. Conclusion

For nanometer sizes, the results (i) for two discernible particles and (ii) taking into account the Pauli principle differ less than room temperature thermal energy kT, which is an adequate precision for microelectronic applications.

With the simple consistent Poisson–Schrödinger solving of interaction between two electrons in a silicon quantum dot, it is possible to define a quantum capacitance. For less-than-six-nanometer dots, this capacitance, in series with the sum of junction and gate capacitance around the dot, increases room temperature Coulomb-blockade effect.

This approach can easily be integrated in a 3D-device-simulator.

References

[1] Grabert H, Devoret M. Single Charge Tunneling: Coulomb Blockade Phenomena in Nanostructures, Series B: Physics Vol. 294, NATO AIS Series, Plenum Press.

[2] Tiwari S, Rana F, Chan K, Shi L, Hanafi H. Single charge and confinement effects in nano-crystal memories. Appl Phys Lett 1996;69(9):1232–4.

[3] Zhuang L, Guo L, Chou SY. Room Temperature Silicon Single-Electron Quantum-Dot Transistor Switch. IEDM Tech Dig 1997;167–169.

[4] Leobandung E, Guo L, Wang Y, Chou SY. Observation of quantum effects and Coulomb blockade in silicon quantum-dot transistors at temperatures over 100 K. Appl Phys Lett 1995;67(7):938–40.

[5] Lamouche G, Fishman G. Two interacting electrons in a three-dimensional parabolic quantum dot: a simple solution, submitted for publication.

[6] Fishman G, Mihalcescu I, Romestain R. Effective-mass approximation and statistical description of luminescence line shape in porous silicon. Phys Rev B 1993;48(3):1464–7.

[7] Sze SM. In: Physics of Semiconductor Devices, 2nd edn. New York: Wiley-Interscience, 1981. p. 397.

Fig. 3. ...

PERGAMON

Solid-State Electronics 43 (1999) 1153–1157

SOLID-STATE ELECTRONICS

Determination of the spectral behaviour of porous silicon based photodiodes

R.J. Martín-Palma*, R. Guerrero-Lemus, J.D. Moreno, J.M. Martínez-Duart

Departamento de Física Aplicada, C-12, Universidad Autónoma de Madrid, 28049 Cantoblanco, Madrid, Spain

Received 19 October 1998; received in revised form 11 January 1999; accepted 27 January 1999

Abstract

Porous silicon (PS) based photodiodes were formed by depositing gold (Au) contacts onto the PS surface. PS was formed from p-silicon substrates under different formation parameters (current density and time of anodization), so PS layers with different porosities and thicknesses were obtained. It was determined the responsivity and the quantum efficiency of these structures in the 200–2500 nm wavelength range, from which it has been observed a different behaviour of those diodes depending on the porosity and thickness of the PS layer. It has also been studied the spectral response from different diodes in which semitransparent conducting films (gold and indium tin oxide) have been deposited onto the PS layer, obtaining a significant improvement in the photoelectronic properties in the visible and near infrared parts of the spectrum. © 1999 Elsevier Science Ltd. All rights reserved.

1. Introduction

Since the discovery of room temperature photoluminescence [1] and electroluminescence [2] in porous silicon (PS), there has been a great interest towards this material due to the possibility of producing optoelectronic devices. Although the research has been mainly focused on the photo and electroluminescent properties, it was found a few years ago that the metal/PS/p-silicon diode structure is light sensitive [3], so PS-based photodetectors can be developed.

In the present work it has been studied the spectral response (responsivity and quantum efficiency) of various PS-based photodiodes as a function of the formation conditions of the PS layer: anodization time and current. The basic structure of these devices is Au/

PS/p-Si/Al. However, it has been also studied the spectral response of photodiodes in which the basic structure was modified by adding semitransparent conducting films (gold and indium tin oxide, ITO) onto the external surface of the PS layer.

2. Experimental

The PS layers were formed by anodization of boron-doped (p-type) silicon wafers of $\langle 100 \rangle$ orientation and with a resistivity of 0.1–0.5 Ω cm. The wafers were back-coated with an aluminium layer ($\sim$2000 Å), were thermally annealed to provide a low resistance ohmic electrical contact and finally were cut into 1.1 × 1.1 cm^2 pieces which were mounted into a sample holder with an exposed area of 0.64 cm^2. The electrolyte consisted of a 2:1 HF (48 wt%):ethanol (98 wt%) mixture. The wafers were galvanostatically etched under illumination. The anodization current density was varied from 40 to 80 mA/cm^2 and so was the time from 180 to 360 s. The samples were immersed in ethanol after

* Corresponding author. Tel.: +34-91-397-4919; fax: +34-91-397-3969.

E-mail address: rauljose.martin@uam.es (R.J. Martín-Palma)

0038-1101/99/$ - see front matter © 1999 Elsevier Science Ltd. All rights reserved.
PII: S0038-1101(99)00039-8

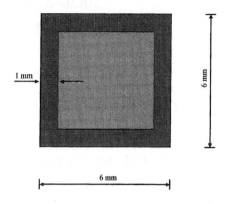

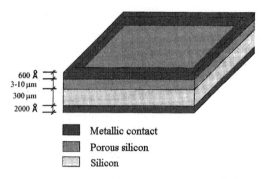

600 Å
3-10 μm
300 μm
2000 Å

■ Metallic contact
▨ Porous silicon
▢ Silicon

Fig. 1. Top and general views of a PS-based photodiode.

the formation of the PS and were immediately loaded into the vacuum chamber for contact deposition. Gold top contacts (~600 Å thick) with the geometry shown in Fig. 1 were sputter-deposited onto the porous silicon layer, so devices with the structure Au/PS/p-Si/Al were obtained. The active area of the resulting photodiodes is of 16 mm². Table 1 summarizes the final structure of the different PS-based devices elaborated. In the case of set of samples D, a semitransparent gold film (~150 Å thick) was deposited onto the PS layer before contact deposition, while in the case of type E samples, an

ITO layer (~900 Å thick) was deposited on top of the PS layer.

For the measurement of the responsivity (R) and the quantum efficiency (η) of the various photodiodes elaborated, these were illuminated by means of an Acton Research Corporation Tungsten–Deuterium dual light source, model TDS-429. It was employed a SpectraPro-150 monochromator, equipped with three interchangeable diffraction gratings, which allow a continuous selection in the 200–2500 nm wavelength range of the radiation employed. To measure the optical power it was employed an International Light radiometer/photometer, model IL1400A. The measurements of the electrical response were taken in the photovoltaic mode, connecting a series resistance to the PS-based photodiode, in which the voltage drop across the resistor and the current through the resistor are governed by Ohm's law, whereas the voltage across the photodiode and the current through the photodiode are governed by the photodiode's photovoltaic characteristic curve [4].

3. Results and discussion

For the purpose of determining the responsivity of the devices, it was measured the voltage drop across the series resistance, from which it was determined the photogenerated current (I_p). The optical power (P_{opt}) was determined by means of a photometer. Therefore, the responsivity, defined as the ratio of the photocurrent to the optical power, can be calculated by using [5]:

$$R(\lambda) = I_p/P_{opt}$$

From this expression, the quantum efficiency (η), i.e. the number of electron–hole pairs generated per incident photon, can be obtained by the equation [6]:

$$\eta = \frac{I_p/q}{P_{opt}/h\nu} = \frac{I_p}{P_{opt}} \frac{h\nu}{q} = R\frac{h\nu}{q}$$

As it can be observed from Fig. 2, the values corresponding to the responsivity are almost zero for wave-

Table 1
Formation conditions and structure of the PS-based photodiodes to be studied

Set of samples	Anodization current (mA/cm²)	Anodization time (s)	Structure
A	40	180	Au/PS/p-silicon/Al
B	80	180	Au/PS/p-silicon/Al
C	40	360	Au/PS/p-silicon/Al
D	40	180	Au/Au(semitransparent)/PS/p-silicon/Al
E	40	180	Au/ITO/PS/p-silicon/Al

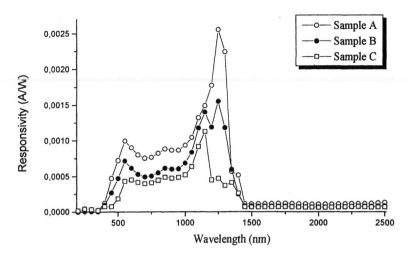

Fig. 2. Spectral responsivity of samples A, B and C.

lengths longer than 1400 nm, which is related to the bandgap energy. The values of the responsivity for wavelengths shorter than 400 nm are also negligible. In this respect, it has to be pointed out that at shorter wavelengths (blue–near UV) the sensitivity is limited by recombination effects near the surface of the semiconductor [7]. It can also be observed from Fig. 2 that PS-based photodiodes show higher responsivity in the near-infrared part of the spectrum. The low values corresponding to the spectral responsivity (of the order of 10^{-3} A/W) are a consequence of working in the photovoltaic mode with a very high value of the series resistance. To obtain higher values of the responsivity, measurements should be taken in the reverse-bias mode to reduce the effect of the diode's internal series resistance [8]. The values of the quantum efficiency (Fig. 3) show two main peaks located in the near-infrared range and in the neighbourhood of 550 nm. The peak located in the visible range can be associated to the top metallic contact since this type of behaviour has been observed in Au–Si photodiodes [6]. The peak corresponding to the infrared range might be associated to the PS/Si interface. It can be also observed in Fig. 3 that the values in the near-infrared range of the quantum efficiency corresponding to type C samples decrease from 1150 nm, while in the case of type A

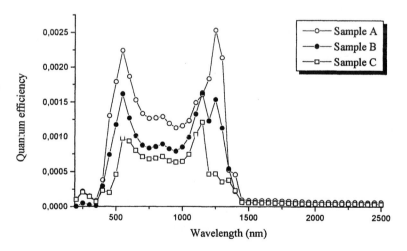

Fig. 3. Spectral quantum efficiency of samples A, B and C.

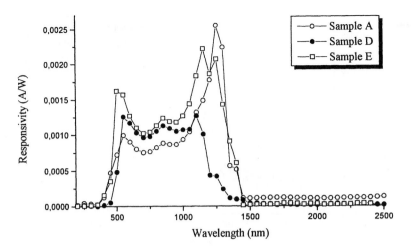

Fig. 4. Spectral responsivity of samples A, D and E.

and B samples, the decrement occurs at longer wavelengths. It has to be noted that type B and C samples present lower values of responsivity and quantum efficiency than those of sample A, which can be associated to a higher series resistance of the PS layer as a consequence of their higher porosity (type B samples) and higher thickness (type C samples), since under unbiased conditions in the photovoltaic mode the load resistance and the internal series resistance of the detector are the most limiting factors [7]. Therefore, it is important to remark the possibility of varying the spectral behaviour of the curves of the responsivity

and the quantum efficiency by varying the formation parameters of the porous layer.

It is clearly observed from Fig. 4 and Fig. 5 that when a semitransparent gold film is deposited onto the PS layer (type D samples), the responsivity as well as the quantum efficiency in the 500–1000 nm wavelength range of this device is higher than the corresponding to type A samples. However, the responsivity and the quantum efficiency are lower for longer wavelengths, which can be explained by the diminution of the transmittance of the gold film for wavelengths shorter than 500 nm and in the near-infrared range [9]. In the case

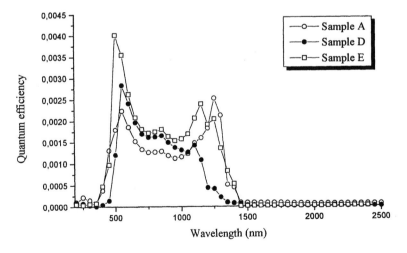

Fig. 5. Spectral quantum efficiency of samples A, D and E.

of type E samples, in which an ITO film is deposited onto the PS layer before contact deposition, the responsivity and the quantum efficiency in the 500–1100 nm wavelength range (Figs. 4 and 5) are notably higher when compared with those of type A samples. For wavelengths higher than 1100 nm, the responsivity is lower than the values of samples A, which can be explained by the decrement of the transmittance of the ITO film for wavelengths above 1100 nm [10]. The values of the responsivity and the quantum efficiency of samples E are also higher than the corresponding to samples D. Therefore, it is obtained an important increment in the responsivity and quantum efficiency in the 500–1000 nm wavelength range when semitransparent conducting films are deposited onto the PS surface. This fact may have its origin in an increment in the collection of photogenerated carriers, although the transmittance of these films limit the photons reaching the surface of the PS layer.

4. Conclusions

It has been determined the spectral behaviour of both responsivity (R) and quantum efficiency (η) of different PS-based photodiodes obtained by a variation of the electrochemical formation parameters of the porous layer. It has been observed that the values corresponding to R and η are negligible for wavelengths longer than 1400 nm and shorter than 400 nm. There exists a significant responsivity in the near-infrared part of the spectrum, decreasing with growing photon energy, except for the peak located in 550 nm. The quantum efficiency spectrum of samples A, B and C show two main peaks, in the near-infrared and in the

visible ranges. It has to be noted that the shape of the curves corresponding to R and η can be varied, within a given limits, by changing the formation parameters of the PS layer. The deposition of semitransparent films (gold or ITO) onto the PS before contact deposition, causes an increment in the values of the responsivity and quantum efficiency measured in the 550–1000 nm range, with respect to the values corresponding to uncoated samples. However, R and η values are limited by the reflectance of the films employed.

References

[1] Canham LT. Appl Phys Lett 1990;57:1046.

[2] Koshida N, Koyama H. Appl Phys Lett 1992;60:347.

[3] Zheng JP, Jiao KL, Shen WP, Anderson WA, Kwok HS. Appl Phys Lett 1992;61:459.

[4] Mooney WJ. Optoelectronic devices and principles. London: Prentice-Hall, 1991 p. 255.

[5] Chuang SL. Physics of optoelectronic devices. New York: Wiley, 1995 p. 599.

[6] Sze SM. In: Physics of semiconductor devices, 2nd ed. New York: Wiley, 1981. p. 749–51.

[7] Angerstein J. UV-visible and near-IR semiconductor sensors. In: Wagner E, Dändliker R, Spenner K, editors. Sensors. Weinheim: VCH, 1992. p. 185–91.

[8] Balagurov LA, Yarkin DG, Petrovicheva GA, Petrova EA, Orlov AF, Andryushin SYa. J Appl Phys 1997;82(9):4647.

[9] Valkonen E, Karlsson B, Ribbing C-G. Solar Energy 1984;32(2):211.

[10] Granqvist CG. Energy-efficient windows: present and forthcoming technology. In: Materials Science for Solar Energy Conversion Systems. Oxford: Pergamon, 1991.

PERGAMON

Solid-State Electronics 43 (1999) 1159–1163

SOLID-STATE ELECTRONICS

The formation of narrow nanocluster bands in Ge-implanted SiO$_2$-layers

J. von Borany[a,*], R. Grötzschel[a], K.-H. Heinig[a], A. Markwitz[a], B. Schmidt[a], W. Skorupa[a], H.-J. Thees[b]

[a]*Forschungszentrum Rossendorf e.V. Institut für Ionenstrahlphysik und Materialforschung, P.O. Box 51 01 19, D-01314 Dresden, Germany*
[b]*Zentrum Mikroelektronik Dresden GmbH, Grenzstraße 28, D-01109 Dresden, Germany*

Abstract

The paper describes the formation of Ge nanocrystals in thin thermally grown SiO$_2$-layers ($d_{ox} \leqslant 100$ nm) using implantation of 10^{15}–2×10^{16} Ge$^+$/cm^2 and subsequent annealing. Although the implanted Ge depth profile is distributed over almost the whole SiO$_2$ layer, a very narrow band (typical width 5 nm) of Ge nanoclusters very close but well-separated to the Si/SiO$_2$-interface is formed by self-organization under specified annealing conditions. A possible mechanisms for this self-organization process is discussed including nucleation phenomena, Ostwald ripening and defect-stimulated interface processes. Simple MOS-structures were prepared and the effect of charge storage inside the clusters has been derived from *C–V* characteristics. © 1999 Elsevier Science Ltd. All rights reserved.

Keywords: Ge nanocrystals; SiO$_2$ films; Ion beam synthesis; Electron microscopy; Non-volatile memory

1. Introduction

Currently, the most popular form of nonvolatile memories (nv-MEM) is flash EEPROM. Large effort has been performed to reduce the programming voltage (12–20 V) of EEPROM's and to overcome their limited endurance ($< 10^6$ write/read cycles). Among different concepts [1,2] the nanocluster memory offers a very promising new idea for nv-MEM's with a programming voltage below 5 V and an expected endurance above 10^9 cycles. The memory cell is designed as a standard field effect transistor (FET), where the gate

oxide contains a Si or Ge nanocrystal layer very close to the Si/SiO$_2$ interface. The nanoclusters (size: 2–5 nm) have to be arranged well separated from each other and with a specified distance of only a few nanometers from the Si/SiO$_2$ interface. The charge exchange between the substrate and the clusters will occur via direct electron tunnelling, leading to a shift in the threshold voltage of the FET with respect to the charge state within the clusters.

So far, two different technologies have been applied to realize such nanocluster containing gate oxides for memory cells. Thin SiO$_2$ films, which have been implanted with Si- or Ge-ions in the range of 10^{15}–10^{16} ions/cm^2, show a significant effect for charge storage in the oxide after annealing [3–9]. However, nanoclusters were not yet observed with electron microscopy and the memory effect is sometimes explained with charge trapping at ion beam induced neutral

* Corresponding author. Tel.: +49-351-260-3378; fax: +49-351-260-3285.
  E-mail address: j.v.borany@fz-rossendorf.de (J. von Borany)

0038-1101/99/$ - see front matter © 1999 Elsevier Science Ltd. All rights reserved.
PII: S 0 0 3 8 - 1 1 0 1 (9 9) 0 0 0 4 0 - 4

traps in the SiO_2 layer [10]. Recently, LPCVD deposition technique has been used to produce self-assembled silicon quantum dots onto ultrathin thermally grown SiO_2 (~3 nm), which are covered by an control oxide of 7–10 nm thickness [11–14]. The electron charging effect of the silicon dots due to direct tunnelling was clearly demonstrated and first memory cells have been realized with an excellent endurance of $> 10^9$ cycles [12]. However, this technology is characterized by a sophisticated multistep procedure of high complexity including two different LPCVD deposition steps of Si and SiO_2 and a local oxidation of the nanoclusters. Contrary, the ion beam synthesis (IBS) of nanoclusters offers a quite simple technology of full compatibility with the present CMOS process. Therefore, IBS should be a favourable technology for the fabrication of nanocluster memory cells, if there is a reliable way to create nanocrystals near the Si/SiO_2 interface, which meet the requirements for nv MEM's.

This contribution describes the formation of narrow bands of crystalline Ge nanoclusters into thermally grown thin SiO_2 layers using ion beam synthesis. The Ge elemental distribution as well as the Ge cluster size and position distribution are investigated in detail by Rutherford backscattering spectroscopy (RBS) and transmission electron microscopy (TEM), respectively. A possible mechanism for the self-organization of near-interface cluster bands is discussed and first measurements at MOS capacitors demonstrate a clear memory effect.

2. Experimental

100 nm SiO_2 layers were thermally grown onto single crystalline Si substrates ([100], n-type, 5–10 Ω cm) in dry oxygen at 1000°C. Subsequently, the gate oxides were implanted with Ge-ions (70–120 keV, 1–2 $\times 10^{16}$ cm^{-2}) up to a Ge concentration in the maximum of the profile between 1 and 5 at%. Post-implantation annealing was performed in the range from 500 up to 1000°C under dry nitrogen in a standard furnace to initiate the Ge nanocluster formation. A set of samples was then separated for microstructural investigations. A Philips CM 300 transmission electron microscope with a line resolution of 0.14 nm was used for TEM analyses of cross-sectional prepared specimens. Additionally, scanning transmission electron microscopy combined with energy dispersive X-ray analysis (STEM-EDX) was applied to determine the depth profiles of Ge, Si and O, simultaneously. The RBS measurements was performed with ^{4}He-ions of 1.7 MeV ion energy.

MOS-capacitors were prepared at a second set of samples to investigate the charge storage properties. The Si wafers were oxidized to 20 nm SiO_2 thickness and subsequently implanted with 20 keV Ge-ions to

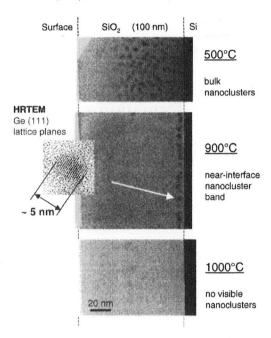

Fig. 1. Sequence of cross-sectional TEM micrographs of Ge implanted SiO_2 films (75 keV, 1.2×10^{16} cm^{-2}) after different annealing treatments (N_2, 60 min). The clusters are characterized by the dark spots, clusters below 2 nm size cannot resolved by the TEM.

fluences in the range of $1–5 \times 10^{15}$ cm^{-2}. Rapid thermal processing (RTP) at 950°C, 30 s under dry nitrogen atmosphere was applied for annealing. The contacts with an area of 0.3 mm^2 were realized by patterned n$^+$-doped poly-Si-dots of 300 nm thickness in the standard technology, which include a 900°C, 10 min thermal treatment step for phosphorous diffusion and a final annealing at 420°C, 10 min under forming gas atmosphere.

3. Microstructure

Fig. 1 shows a sequence of cross sectional TEM micrographs of SiO_2 films implanted with 75 keV Ge ions (1.2×10^{16} cm^{-2}) for different annealing temperatures. After 500°C annealing a region of amorphous Ge nanoclusters is found in the oxide bulk around the maximum of the as-implanted profile. After an annealing step of 900°C no bulk nanoclusters are further visible, but a narrow nanocluster band appears very close but well separated to the Si/SiO_2 interface. The mean size of the nanoclusters inside the band is about 4 nm and a cluster density of around 2×10^{12} cm^{-2} can be determined. Using high resolution TEM, the Ge (111)

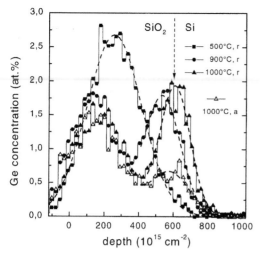

Fig. 2. Ge depth profiles measured by RBS. The closed symbols show the spectra for random incidence, the open triangle represents the result for 1000°C annealing temperature under channelling conditions.

lattice planes of single nanoclusters have been found showing the crystalline state of the clusters. At higher annealing temperatures this near-interface nanocluster band disappears, which confirms the transient character of the nanocluster solution.

The Ge depth profiles determined by RBS are summarized in Fig. 2. The Ge distribution after 500°C annealing corresponds to the as implanted profile ($R_p = 56$ nm, $\Delta R_p = 18$ nm). After annealing at temperatures ≥ 900°C a significant redistribution of Ge has been found, which is characterized by an accumulation of Ge near the Si/SiO$_2$ interface of nearly half of the whole Ge content implanted into the oxide.[1] With increasing temperature the Ge peak position is slightly shifted towards the Si substrate, but no significant change in the Ge fraction of the near interface peak is obtained. The shift in the peak position can be explained from STEM-EDX results (Fig. 3). With this method the surface and the Si/SiO$_2$ interface can be determined with a high accuracy (<2 nm) from the peak positions of the differentiated Si signals. It is clearly shown that after annealing at 900°C the Ge peak is located inside the SiO$_2$ layer and corresponds with the position of the nanocluster band observed by TEM. Contrary, after annealing at 1000°C most of the

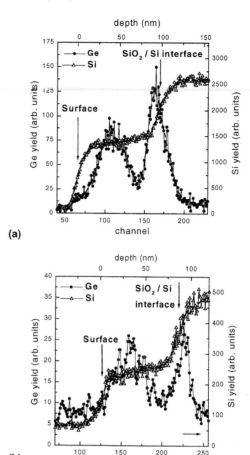

(a)

(b)

Fig. 3. Ge and Si line scan profiles across the SiO$_2$ film obtained by STEM-EDX for specimens annealed at 900°C (a) and 1000°C (b), respectively.

near-interface Ge is incorporated in a small layer in the top of the Si substrate and no nanoclusters in the SiO$_2$ can be found. From the reduced Ge yield of the RBS spectrum under aligned incidence (Fig. 2: 1000°C, a) referring to the random case one can conclude, that most of the near-interface Ge is incorporated into the silicon lattice after annealing at 1000°C.

The 'thermal window' for the existence of the interface nanocluster band can be influenced by the implantation and annealing conditions (e.g. atmosphere, annealing time), which determine the amount of available (non-bonded) Ge. For a similar experiment using slightly larger implantation energy and fluence a nanocluster band can be observed after annealing at 1000°C (Fig. 4). Under these conditions additional crystalline nanoclusters exist in the SiO$_2$ bulk and the

---

[1] The Ge accumulation at the Si/SiO$_2$ interface has been found for nearly all cases of Ge implanted SiO$_2$ layers after annealing at elevated temperatures (>900°C), even for the case that the implantation profile is far away from the interface.

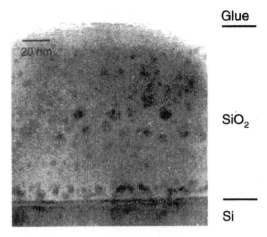

Fig. 4. XTEM micrograph of a Ge implanted SiO_2 films (110 keV, 2×10^{16} cm^{-2}) after annealing at 1000°C, N_2, 60 min. Besides the near-interface nanocluster band, additional nanoclusters are clearly visible in the SiO_2.

interface nanocluster band and the bulk nanoclusters are separated by a denuded zone without visible clusters.

4. Origin of the nanocluster layer at the Si/SiO$_2$ interface

A controlled fabrication of the narrow near-interface nanocluster layer can be facilitated by a basic understanding of the mechanisms involved. What is the reason for the evolution of a separated near-interface nanocluster band, which position differs to the implanted Ge profile?

Referring to our experiment, the Ge concentration in the region of the narrow nanocluster band amounts to only 1×10^{20} cm^{-3} (0.15 at%) as determined by TRIM calculation. Therefore, diffusion of Ge from the implanted SiO_2 region and an accumulation of Ge at nucleation centers in a thin layer parallel to the Si/SiO_2 interface have to be included in the model of nanocluster band formation. These nucleation centers are assumed to be small Si agglomerates which are formed as a result of defect-stimulated interface processes as it is shown elsewhere in more detail [15]. TRIM calculations show, that during Ge ion implantation (75 keV, 1.2×10^{16} cm^{-2}) into 100 nm SiO_2 collisional mixing mainly by the recoil atoms leads to a dissociation of the oxide into its elemental components O and Si by. Even at the interface the atoms of SiO_2 will be displaced several times. Although most of the displaced Si and O atoms react very fast to recover the SiO_2, a small steady state fraction of non-bonded Si and O will be present during ion implantation. In the course of studies of the SIMOX process it has been shown, that non-bonded oxygen atoms in SiO_2 diffuse very fast during ion implantation even at room temperature. As the SiO_2/Si interface is a sink for diffusing non-bonded oxygen, there will evolve a narrow steady-state zone of SiO_2 with oxygen deficiency. In this zone there is no longer sufficient non-bonded oxygen for the

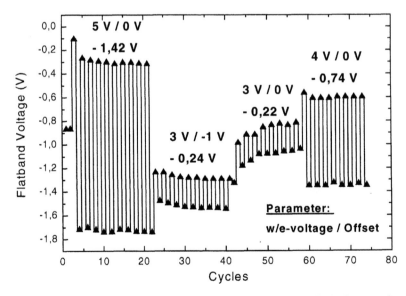

Fig. 5. Shift of the flatband voltage (V_{fb}) after erase/write cycling with different pulse amplitudes (programming time: 10 ms).

recombination with non-bonded silicon, which is produced at the same rate by collisional mixing. Thus, the concentration of non-bonded silicon increases in the near-interface region. At the end of the ion implantation process, there remains a silicon excess in the SiO_2 at the interface, which forms Si agglomerates in the subsequent annealing step and act as nucleation centers for diffusing Ge from the implanted SiO_2 regions.

Up to now, this model explains for all performed experiments why this nanocluster band evolves or why it not evolves. A prerequisite for the nanocluster band formation is a total implantation damage at the interface of at least 1 displacements per target atom. Thus, if the Ge energy or the Ge fluence are too low in order to produce sufficient damage at the interface, no nanocluster band is formed. It has been proven, that the formation of the near-interface nanocluster band is also avoided for sufficient damage at the interface or if the mobility of the non-bonded oxygen is low (e.g. during implantation at liquid nitrogen temperature).

5. Memory effect

MOS capacitors containing nanocrystals in the gate-oxide were used to investigate the feasibility for a memory device. When forward biasing the gate with respect to the substrate electrons are injected into the nanocrystals and cause a positive threshold shift (write cycle). The charge can be removed by a programming pulse of opposite polarity (erase cycle). Fig. 5 shows some results of high frequency capacitance–voltage (C–V) measurements on a Ge-implanted sample (5×10^{15} cm^2). In this example erase and write pulses of different amplitudes were applied to the structure while the programming time per pulse was kept constant at 10 ms. From the high frequency C–V curves the shift of the flatband voltage for 3, 4 and 5 V pulses was measured and determined to be -0.22, -0.74 and -1.42 V, respectively. Note that by using current sensing for the read operation of the memory even a programming window of -0.22 V may be sufficient. In summary a clear memory effect could be demonstrated. Further investigations on all aspects of memory operation (retention, endurance etc.) are under way.

6. Conclusions

Cross sectional TEM investigations show the ability to form narrow semiconductor nanocluster bands at the Si/SiO_2 interface after ion implantation of Ge ions into thin SiO_2 layers ($\leqslant 100$ nm) and subsequent annealing. The band of separated (crystalline) nanoclusters is located only few nanometers from the Si/SiO_2 interface and fulfill very well the requirements for memory applications. The evolution of nanoclusters is a transient process and the existence of the near interface nanocluster band can only be observed under specific implantation and annealing conditions. The 'thermal window' depends from the availability of non-bonded Ge in the SiO_2 film, which can diffuse towards the Si/SiO_2 interface and segregate at nucleation centers. According to our model, this nucleation centers are formed by excess silicon near the Si/SiO_2 interface, which arises as a result of damage and recovering of the SiO_2 network during ion implantation. Flatband voltage shifts of > 200 mV for programming pulses < 5 V ($t = 10$ ms) has been obtained for Ge implanted 20 nm SiO_2 layers ($d_{ox} = 20$ nm) which can be attributed to the electron exchange between the clusters and the substrate. Whether the charge is stored in the clusters or at cluster related deep trapping centers, as it is concluded from long tome retention investigations for a silicon nanocrystal MOS-memory [16] remain still an open question.

References

[1] Ahmed H. J Vac Sci Technol B 1997;15(6):2101.

[2] Eitan B. Microelectron Eng 1997;36:277.

[3] Hori T, Ohzone T, Odark Y, Hirase J. Tech Dig 1992:469.

[4] Hao M, Hwang H, Lee JC. Appl Phys Lett 1993;62:1530.

[5] Hao M, Hwang H, Lee JC. Solid-St Electron 1993;36:1321.

[6] Ohzone T, Hori T. Solid-St Electron 1994;37:1771.

[7] Hanafi H, Tiwari S. In: ESSDERC '95 Tech. Dig., The Hague, The Netherlands, 25–27 September, 1995. p. 209.

[8] Hanafi H, Tiwari S, Khan I. IEEE Trans Electron Devel 1996;43(9):1553.

[9] Ohzone T, Matsuda T, Hori T. IEEE Trans Electron Devel 1996;43(9):1375.

[10] Kalnitzky A, Boothroyd AR, Ellul JP. Tech Dig 1988:516.

[11] Maiti B, Lee JC. Electron Device Lett 1992;13(12):624.

[12] Tiwari S, Rana F, Hanafi H, Hartstein A, Crabbe EF, Chan K. Appl Phys Lett 1996;68:1377.

[13] Tiwari S, Rana F, Chan K, Shi L, Hanafi H. Appl Phys Lett 1996;69:1232.

[14] Kohno A, Marakami H, Ikeda M, Miyazaki S, Hirose M. In: Extended abstract of the International Conference on Solid State Devices and Materials, Hamamatsu, 1997. p. 566.

[15] Heinig KH, Schmidt B, Markwitz A, Grötzschel R, Strobel M, von Borany J. Nanocrystal formation in SiO$_2$: experiments, modelling and computer simulations, EMRS-1998, 1998 (session J-X1.1, invited).

[16] Shi Y, Saito K, Ishikuro H, Hiramoto T. J Appl Phys 1998;84(4):2358.

PERGAMON

Solid-State Electronics 43 (1999) 1165–1169

SOLID-STATE ELECTRONICS

Influence of carbon-containing contamination in the luminescence of porous silicon

R. Guerrero-Lemus [a],*, F.A. Ben Hander [a], J.D. Moreno [a], R.J. Martín-Palma [a], J.M. Martínez-Duart [a], P. Gómez-Garrido [b]

[a]*Departamento de Fisica Aplicada C-12 and Instituto de Ciencia de Materiales, CSIC, Universidad Autónoma de Madrid, 28049 Madrid, Spain*
[b]*Departamento de Física Fundamental y Experimental, Universidad de La Laguna, 38206 La Laguna, S/C de Tenerife, Spain*

Received 31 July 1998; received in revised form 9 January 1999; accepted 27 January 1999

Abstract

This paper is focused to study the variations of the photoluminescent properties of porous silicon when samples are immersed in deionized water or ethanol after etching. The emission peaks in the photoluminescence spectrum redshift and decreases with immersion time. We have observed that the photoluminescence disappears after about 20 h immersion in ethanol. The results are interpreted in terms of carbon contamination in the surface. This hypothesis is confirmed by Fourier transform infrared and X-ray photoelectron spectroscopies. © 1999 Published by Elsevier Science Ltd. All rights reserved.

1. Introduction

Research on porous silicon (PS), grown on a Si substrate by anodic electrochemical etching in a HF solution, is presently of great interest [1–3]. The recent observation that these porous layers exhibit relatively intense visible photoluminescence at room temperature [4] has attracted considerable attention and provided encouraging prospects for Si-based optoelectronic integrated circuits [5]. However, the detailed mechanism of the luminescence, characterized by a broad Gaussian-like peak in the visible region, remains relatively unknown.

This work is focused to the study of the photoluminescence evolution of PS when the samples are immersed in deionized water or ethanol, as well as the role of surface passivants in the mechanism of photoluminescence. For this purpose we have obtained the emission, infrared and X-ray photoelectron spectra of the different samples analyzed.

2. Experimental procedure

The samples were boron doped (0.1–0.5 Ω·cm) silicon wafers (100), electrochemically prepared [6] under illumination in a 1:1 HF (48 wt%)/ethanol (98 wt%) solution during 15 min at a current density of 20 mA/cm^2. The samples were measured after being rinsed in deionized water or in ethanol and after 1 or 23 h immersed in deionized water and ethanol.

The electrochemical parameters were controlled by a computer-driven electrochemistry system (PAR model 273). Emission spectra were acquired with an AMINCO-Bowman series 2 luminescence computer-controlled spectrometer and the excitation wavelength

* Corresponding author. Tel.: +34-91-397-4918; fax: +34-91-397-3969.
E-mail address: ricardo.guerrero@uam.es (R. Guerrero-Lemus)

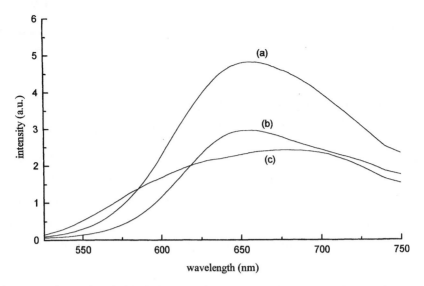

Fig. 1. Emission spectra of PS (a) rinsed in deionized water; (b) immersed in deionized water for 1 h; and (c) immersed in deionized water for 23 h.

was fixed at 400 nm. Fourier transform infrared (FTIR) spectra were acquired in order to determine the chemical composition of the porous surface following the different post etch treatments. Specular reflectance Fourier transform infrared spectra were recorded with a Bruker IFS 66 V computer controlled spec-

trometer. Photoelectron spectra were obtained using a Fisons Escalab 200R spectrometer equipped with a hemispherical electron analyzer and a MgKα ($h\cdot\nu = 1253.6$ eV) 120 W X-ray source. High resolution spectra were obtained at a pass energy of 20 eV by

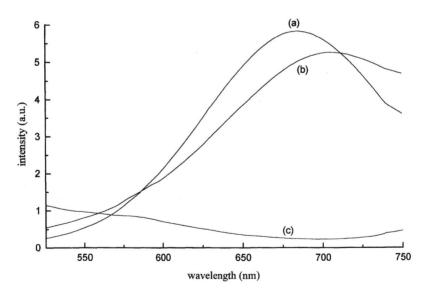

Fig. 2. Emission spectra of PS (a) rinsed in ethanol; (b) immersed in ethanol for 1 h; and (c) immersed in ethanol for 23 h.

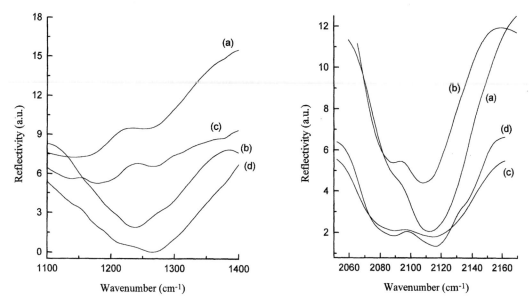

Fig. 3. FTIR spectra of the Si–CH₃ absorption band at 1245–1275 cm⁻¹ for PS (a) immersed in deionized water 1 h; (b) immersed in ethanol 1 h; (c) immersed in deionized water 23 h; and (d) immersed in ethanol 23 h.

Fig. 4. FTIR spectra of the Si–H absorption band at 2080–2130 cm⁻¹ for PS (a) immersed in deionized water 1 h; (b) immersed in ethanol 1 h; (c) immersed in deionized water 23 h; and (d) immersed in ethanol 23 h.

scanning each energy region several times in order to obtain good signal-to-noise ratios.

3. Results and discussions

In Fig. 1 the emission spectra of PS rinsed in deionized water and immersed in deionized water for 1 or 23 h are shown. The three peaks are located at 654, 657 and 680 nm, respectively. The intensity of the emission peaks is decreasing with the immersion time, and the widths of the peaks are significantly increasing during the process.

Fig. 2 shows the emission spectra of PS rinsed in ethanol and immersed in ethanol 1 or 23 h after the etching. The peak for the rinsed sample is situated at 685 and 705 nm for the sample immersed 1 h in ethanol. However, the light signal for the sample immersed 23 h in ethanol is negligible.

Fig. 3 shows the FTIR spectra of PS samples corresponding to the peaks located at 1245–1275 cm⁻¹ which are attributed to Si–CH₃ bonds [7]. It can be appreciated that only for samples immersed in ethanol this peak is appreciable, and the magnitude of the peak increase with the immersion time. For samples immersed in deionized water, only small traces of the SiCH₃ absorption band are detected. The width of the absorption peaks, higher than usual [7], are associated

to the special characteristics of the porous structures, with stresses and disordered Si–CH₃ bonds, which introduces a relative dispersion of the vibrations around the central peak.

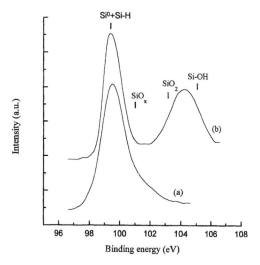

Fig. 5. Si(2p) core level spectra of just formed PS samples: (a) followed by a rinse in deionized water; (c) id. followed by a rinse in ethanol.

Table 1
Percentage of elements in PS after different postetch treatments

Sample	Si (%)	O (%)	C (%)	F (%)
PS rinsed in deionized water	76	20	2	2
PS rinsed in ethanol	26	41	29	4

Fig. 4 shows the FTIR spectra of the Si–H stretching bands of PS samples, situated at 2080–2130 cm^{-1} [8]. It can be appreciated a decrease of the magnitude of this peak for the two kinds of samples, but this evolution is more significant for the samples immersed in deionized water. This figure represents the change of the porous silicon surface passivation from Si–H when the sample is just formed to Si–O or Si–CH$_3$ depending of the solvent composition used.

The two types of samples rinsed in deionized water or ethanol have been also studied by XPS in order to determine their surface composition. The energy regions corresponding to C(1s), O(1s), Si(2p) and F(1s) were recorded for each sample. The intensity value of the peak associated to N atoms was negligible, as well as the C(1s) peak corresponding to samples rinsed in deionized water. The XPS results indicate that the most relevant differences between samples occur in the Si(2p) profile (Fig. 5), where the increase of the peak widths with respect to monocrystalline silicon denotes an important presence of Si–H bonds and some amorphization [9]. PS rinsed in deionized water (line a) contains the contribution of a suboxide at higher binding energies (~101 eV). The PS rinsed in ethanol spectrum (line b) shows only small traces of suboxide contribution; however, a high amount of Si–OH bonds is detected [10]. The quantitative contribution of the different elements detected by XPS is shown in Table 1. From these results, and those obtained by FTIR, it can be inferred that PS samples rinsed in deionized water are oxidized and hydrogen passivated, and PS samples rinsed in ethanol are hydrogen and hydroxide passivated.

The differences between emission spectra (maxima and intensity values) for PS immersed in deionized water and ethanol can be explained in terms of the quality of the oxide passivation and carbon-related compounds present in each sample. PS immersion in aqueous solutions is expected to produce lower quality silicon oxides in PS samples than in the case in which the oxidation process is induced, like thermal oxidation or immersion in CuCl$_2$ [11].

In the evolution study of the emission peak of the different samples, a redshift is observed. This behavior can be explained in terms of the quantum confinement model [1] influenced by surface states [12], in which

emission centers are modeled as silicon nanospheres embedded in a fibrilar matrix. The emission efficiency depends of the number of nonradiative centers situated on the nanospheres. For just formed samples, the PS surface is initially passivated by hydrogen, but this passivation is not stable in contact to the solvent, as deduced from Fig. 4. Thus, when the PS structures are being oxidized in the solvent, it is easier for oxygen to access to silicon nanospheres when the surface is initially hydrogen passivated than when is oxidized by more qualified methods. This poor qualified oxidation could induce the formation of surface defects on the silicon nanosperes which reduces the photoluminescent intensity and increases the dispersion of the emission center wavelengths.

Also, the light emission from samples in contact with deionized water show a peak maximum at lower wavelengths than samples in contact with ethanol. This result can be explained from the fact that PS in contact with deionized water is more intensively oxidized than samples in contact with ethanol. Thus, as oxidation reduces the volume of Si in the nanosphere, the emission peak is blueshifted for samples oxidized with respect to non oxidized samples.

From the intensity evolution of the emission spectra for samples immersed in ethanol (Fig. 2) and the PS surface composition (Figs. 3 and 4) it can be inferred that there exists a strong influence of Si–CH$_3$ species, but this C contamination does not reduce the volume of the silicon nanospheres as oxidation does. However, organic compounds like ethanol solvents seem to induce nonradiative traps in the PS surface that dramatically reduce the intensity of the luminescence [13]. Since samples immersed in ethanol show an appreciable carbon related contamination (Fig. 3 and Table 1), which increases with the time immersion, we attribute the extinction of photoluminescence to carbon contamination.

4. Conclusions

In conclusion, the photoluminescent emission spectra evolution for porous silicon samples immersed in deionized water and ethanol shows a small redshift, an increase of the width of the spectra and a decrease of the luminescence intensity. For the case of samples immersed in ethanol, the photoluminescence even disappears after a given time. This is attributed to the carbon-related contamination induced in these samples. We have interpreted the photoluminescent evolution in terms of the quantum confinement model influenced by surface states.

Acknowledgements

We wish to thank Professor J. Piqueras for his assistance with the infrared measurements. We are also indebted to the Spanish CICyT (Project No. MAT96-0602) for the financial help to carry out this research.

References

[1] Cullis AG, Canham LT. Nature 1991;353:335.

[2] Martínez-Duart JM, Parkhutik VP, Guerrero-Lemus R, Moreno JD. Adv Mater 1995;7:226.

[3] Collins RT, Fauchet PM, Tischler MA. Phys Today 1997;50:24.

[4] Canham LT. Appl Phys Lett 1990;57:1047.

[5] Koshida N, Koyama H. Appl Phys Lett 1992;60:314.

[6] Guerrero-Lemus R, Moreno JD, Martínez-Duart JM, Corral JL. Rev Sci Instrum 1996;67:3627.

[7] Launer PJ. Silicon compounds register and review. Petrarch Systems Silanes and Silicones, Petrarch Systems, 1987.

[8] Gupta P, Covin VL, George SM. Phys Rev B 1988;37:8234.

[9] Ley L, Reichardt J, Johnson R-L. Phys Rev Lett 1982;49:1664.

[10] Guerrero-Lemus R, Fierro JLG, Moreno JD, Martínez-Duart JM. Mater Sci Technol 1995;11:711.

[11] Rigakis N, Hilliard J, Abu Hassan L, Hetrick JM, Andsager D, Nayfeh MH. J Appl Phys 1997;81:440.

[12] Koch F. Mat Res Soc Symp Proc 1993;298:319.

[13] Guerrero-Lemus R, Moreno JD, Martínez-Duart JM, Fierro JLG, Gómez-Garrido P. Thin Sol Films, in press.

PERGAMON

Solid-State Electronics 43 (1999) 1171–1175

SOLID-STATE
ELECTRONICS

CdS doped-MOR type zeolite characterization

H. Villavicencio García[a], M. Hernández Vélez[a], O. Sánchez Garrido[b],
J.M. Martínez Duart[c],*, J. Jiménez[d]

[a]*Grupo de Zeolitas y Propiedades Dieléctricas en Sólidos, I.S.P. "E.J. Varona". C. Libertad, Marianao, C. de la Habana, Cuba*
[b]*Instituto de Ciencia de Materiales, CSIC, Madrid, Spain*
[c]*Departamento de Física Aplicada C-XII, Instituto de Ciencia de Materiales, UAM, Cantoblanco, 28049 Madrid, Spain*
[d]*Departamento de Física de la Materia Condensada, Universidad de Valladolid, Valladolid, Spain*

Received 15 September 1998; received in revised form 11 January 1999; accepted 26 January 1999

Abstract

A preliminary characterization of CdS particles grown into zeolite cavities is reported. We have obtained via hydrothermal synthesis a synthetic mordenite type zeolite (host) with CdS particles inside its porous structure. The zeolites were doped with CdS at different interchange ratios. The composite materials obtained were characterized by Raman spectroscopy, optical absorption in the visible range and N_2 physical adsorption. The results allow us to confirm the partial amorphization of the starting zeolitic structures and the presence of semiconducting inclusions in their pores and cavities. The optical properties of the samples showed quantum confinement effects. © 1999 Elsevier Science Ltd. All rights reserved.

1. Introduction

In the second half of this century, new zeolitic materials doped with polymers and metals have been obtained and studied because of their interesting properties to applications in many industrial processes, such as: selective electrodes, chemical sensors, molecular sieves, batteries, non-linear optical materials, quantum well devices and solar energy converters [1–3]. At the same time, the theoretical and experimental researches in the semiconducting materials field have been directed towards the study and synthesis of semiconductor nanostructures deposited into different kind of matrices [4,5]. The physical and chemical properties of the zeolites, which make them very attractive as a host matrix for semiconductor inclusions, have been summarized by Wang and Herron [6].

In this work we used the method of chemical hydrothermal synthesis to grow CdS particles inside the zeolite channels. The obtained results from Raman and visible absorption spectroscopies, as well as the N_2 physical adsorption experiments, confirm the potentiality of the hydrothermal technique and the convenience to improve it in order to obtain desirable quantum effects at atomic scale.

2. Experimental

MOR type zeolites were synthesized hydrothermally following standard procedures reported in the literature [7]. The most accepted stoichiometric composition [8] for this type of zeolite is: $Na_8[Al_8Si_{40}O_{96}]\cdot24H_2O$. It presents an orthorhombic crystalline structure with twelve member channels in the [001] direction and eight member channels in the [010] direction which

* Corresponding author. Fax: +34-91-3973969.
  E-mail address: martinez.duart@uam.es (J.M. Martínez Duart)

Table 1
N$_2$ adsorption results obtained for samples M and M8

Sample	M	M8
Superficial area (BET) (m^2/g)	315.76	10.24
Superficial area. (Langmuir) (m^2/g)	439.07	14.23
Micropore area (m^2/g)	295.496	10.28
Micropore volume (cc/g)	0.146	0.005
Pore mean diameter Langmuir (nm)	1.59	3.82

Table 2
Main Raman parameters of the samples in the 200–700 cm^{-1} range

Sample	Raman shift (cm^{-1})	FWMH (cm^{-1})	I (a.u.)
M	328	43.9	128
	405	35.3	771
	451	34.8	467
	475	23.4	590
M8	308	27.3	203
	406	43.6	375
	460	44.4	379
	486	18.0	467
	611	29.2	119
M11	305	20.8	937
	604	34.5	538
M13	304	25.3	858
	603	33.1	487

have apertures of 6.5 × 7.0 Å and 2.6 × 5.7 Å respectively.

The as-synthesized product was carried out to acid form using a 0.1 M solution of NH$_4$NO$_3$, which will be labeled M. To carry out the ionic exchange of the sample M with cadmium, four solutions containing 10 g of Cd(NO$_3$)$_2$, 0.32 M, in 10, 20, 30 and 40 ml of bidestilled water were prepared obtaining samples labeled: M8, M9, M11 and M13, respectively, in which the exchange ratio increases from M8 to M15. After this, the samples were washed and dried and finally they were submitted to a constant sulphide flow during one hour in a dry atmosphere.

Raman spectra were obtained by means of a Dylor XY spectrometer with an argon laser (line 488 nm). The visible absorption spectra were obtained using an Hitachi 150–20 spectrophotometer with an integrating sphere in the 300–900 nm range. The nitrogen adsorption isotherms were obtained automatically following a volumetric technique.

3. Results and discussion

The results obtained from N$_2$ adsorption exper-

iments for the matrix M and the sample M8, are shown in Table 1. These parameters were calculated from the adsorption isotherms following the BET and Langmuir models [7]. These results confirm the presence of CdS particles in the zeolite channels and cavities that could show quantum confinement effects. At the same time the pore mean diameter evolution allow us to confirm the losses of crystallinity as the Cd concentration increases. This can be understood by the deformation of the pore geometry as it is revealed by adsorption data. The decrease of the micropore volume and area was already noticeable for sample M8.

Fig. 1 shows the pore volume behavior versus mean pore diameter in samples M and M8. Both samples have a relative maximum in a mean pore diameter value near to the smaller cavity dimension, characteristic of the mordenite type zeolite, although in sample

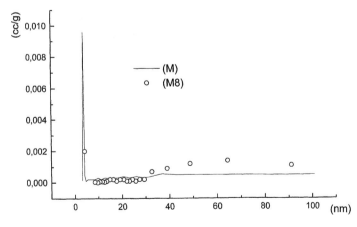

Fig. 1. Pore volume versus pore mean diameter of the samples M and M8.

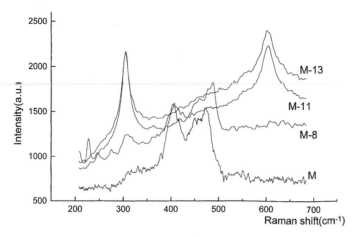

Fig. 2. Raman shift of the samples from 200 to 700 cm^{-1}. All the spectra are shifted vertically for clarity.

M8, the volume corresponding to that dimension is an order of magnitude smaller than in sample M. On the other hand, in sample M8, there exists a large pore size distribution centered about 60 nm, which is in agreement with the structural changes occurring in the insulator matrix when doped with the semiconductor. Evidently the volume of free space significantly decrease from samples M to M8.

The Raman spectra obtained for all types of samples are shown in Figs. 2 and 3 for two different spectral ranges. The main Raman parameters are shown in Table 2. Zeolites show several Raman structures between 200 and 700 cm^{-1}, while hexagonal CdS exhibits a finest order Raman band at 304 cm^{-1} and a second order one around 600 cm^{-1}, which correspond to

the first order (LO) and second order (2LO) vibrations of hexagonal CdS respectively [9,10].

A band at 328 cm^{-1} was only observed in the as-grown sample (M), and can be associated with a framework vibration. The other bands above 400 cm^{-1} are due to the motion of oxygen atoms in a plane perpendicular to the Si(Al)–O–Si(Al) bonds. Samples with Cd inclusions exhibit Raman bands above 300 and 600 cm^{-1}, which are associated with the formation of CdS particles in the zeolite host. The evolution of the Raman spectrum reveals also the progressive loss of crystallinity of the zeolite as more CdS is incorporated, though the structure was not distorted enough to become amorphous at the compositions herein involved. The main Raman bands of the mordenite are

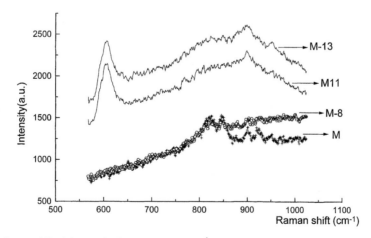

Fig. 3. Raman shift of the samples from 500 to 1100 cm^{-1}. All the spectra are shifted vertically for clarity.

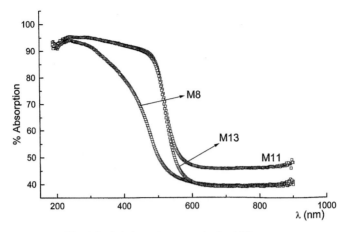

Fig. 4. Optical absorption spectra in the visible range.

always well defined though their intensities progressively decrease.

When the CdS concentration increases, the Raman spectra strongly reflect the CdS particles growth. The crystallite quality in M11 is better than in M8, but it is worst in M13. This behavior suggests that there exists a critical value of Cd content to optimize the CdS particles growth inside the zeolite cavities. The bands observed around 600 cm^{-1}, can be associated to a second order of the LO mode, which could be understood either as a resonant Raman effect, due to the limited size of the CdS particles or as transitions from the cluster system to a supercluster system [11]. More experimental data are necessary to confirm one of these hypothesis. The I_2/I_1 ratios in samples M11 and M13, are approximately 0.57, indicating in a first approach the same crystallite sizes in both samples as we mentioned above.

The optical absorption spectra of the doped samples from 300 to 900 nm are shown in Fig. 4. In these spectra we can see a shoulder with excitonic shape in sample M8. Similarly, from these spectra we have calculated the band gap of these materials obtaining 3.4, 2.63 and 2.59 eV, for samples M8, M11 and M13 respectively, being the last one very near to the band gap of bulk CdS. It is evident that the samples M11 and M13, were strongly doped and there is not a great difference between their optical behavior in the visible range. Once again this fact points out the existence of some Cd ionic exchange limiting the final CdS concentration obtained in the porous structure of the zeolite.

In addition, the obtained value for the band gap of the sample M8, reflects a blue shifting in the emission spectra for the lowest CdS concentration inside the cavities and channels of these porous structures [12]. The largest band gap obtained for this sample is in

agreement with the smallest particle size obtained from the Raman spectra. The last result is similar to that obtained by Wang and Herron for CdS clusters grown in Y-type zeolite [6].

4. Conclusions

CdS nanoparticles smaller than 60 nm were obtained in the interior of mordenite type zeolites. The doping of the matrix via ionic exchange indicates some crystallinity losses although, even the most heavily doped sample shows structural elements corresponding to the starting mordenite. The highest blue shifting of the band gap was obtained for the lowest CdS concentration. The final product constitutes a composite material, formed by a matrix with larger cavities than mordenite filled by the CdS nanoparticles. The results point out also, the existence of a CdS concentration limiting the obtention of nanoparticles in this kind of porous materials.

Acknowledgements

The authors would like to acknowledge the Ministry of Spanish Education for financial support under the Project: Optoelectronic Properties of Nanostructured Materials (1997).

References

[1] Ozin GA, et al. Angew Chem Int Ed Engl 1989;28(3):359.
[2] Chao TH, Erf HA. Journal Catal 1986;100:492.
[3] Thomas JM. In: Jacobs PA, van Santen RA, editors.

Zeolites: Facts, Figures, Future. Amsterdam: Elsevier, 1989. p. 3.

[4] Heuglin A. Top Current Chem 1988;143:113.

[5] Borrelli NF, et al. J Appl Phys 1987;61:5399.

[6] Wang Y, Herron N. J Phys Chem 1987;91:257.

[7] Roque-Malherbe R. Físico-Química de Zeolitas, Monography. Havana, Cuba: Ed. EMPES, 1989.

[8] Meier WM, Olson DH. Atlas of Zeolite Structure Types. London: Butterworth-Heinemann, 1992.

[9] Pechar F, Rykl D. Zeolites 1983;3:329.

[10] Landot B. In: Numerical Data and Functional Relationships in Science and Technology, New series group III, V. 17. Semiconductor. Berlin: Springer Verlag, 1988.

[11] Persons PD, Silvestri M, Mei G, Lu E, Yukselici H, Schroeder J. Brazilian Journal of Physics 1993;23(2):144.

[12] Yoffe AD. Advances in Phys 1993;42(2):173–266.

Printed and bound by CPI Group (UK) Ltd, Croydon, CR0 4YY

08/05/2025

01864849-0004